A Peace in Southern Africa: The Lancaster House Conference on Rhodesia, 1979

Also of Interest

Crisis in Africa: Battleground of East and West, Arthur Gavshon

Foreign Intervention in Civil War: The Politics and Diplomacy of the Angolan Conflict, Charles Kurtz Ebinger

Social Conflicts and Third Parties: Strategies of Conflict Resolution, Jacob Bercovitch

U.S. Economic Power and Political Influence in Namibia, 1700–1982, Allan D. Cooper

†*A History of South Africa to 1870,* edited by Monica Wilson and Leonard Thompson

South Africa into the 1980s, Richard E. Bissell and Chester A. Crocker

The Angolan War: A Study in Soviet Policy in the Third World, Arthur Jay Klinghoffer

†*Alternative Futures for Africa,* edited by Timothy M. Shaw

PROFILES OF CONTEMPORARY AFRICA:

†*Mozambique: From Colonialism to Revolution, 1900–1982,* Allen Isaacman and Barbara Isaacman

†*Tanzania: An African Experiment,* Rodger Yeager

Swaziland: Tradition and Change in a Southern African Kingdom, Alan R. Booth

Botswana: Liberal Democracy and the Labor Reserve in Southern Africa, Jack Parson

†*Kenya: The Quest for Prosperity,* Norman N. Miller

†Available in hardcover and paperback.

Westview Special Studies on Africa

A Peace in Southern Africa: The Lancaster House Conference on Rhodesia, 1979
Jeffrey Davidow

In the middle of 1979 Rhodesia formed one leg of the triad of southern Africa's remaining white-ruled states. The country appeared no closer to peace and majority rule than it had at any time since Ian Smith's 1965 Unilateral Declaration of Independence. But by the end of that year a remarkable agreement had been forged that ended Rhodesia's rebellion and ushered in Zimbabwe's birth.

A Peace in Southern Africa details the personalities, the events, and the negotiating tactics of the Lancaster House Conference that brought peace to Rhodesia/Zimbabwe. It also offers some thoughts on the possible roles of powerful third parties in resolving regional conflicts, an issue of particular relevance to the United States.

Mr. Davidow was able to observe the events surrounding the creation of Zimbabwe from a unique vantage point. A career Foreign Service officer working with southern African affairs during most of the last decade, he was in 1979 the first U.S. official to be posted in Rhodesia since the withdrawal of U.S. officials from that country in 1970. During the 1982–1983 academic year, Mr. Davidow was a fellow of the Center for International Affairs at Harvard University.

"It is superb, a rich tapestry, done with style and verve and some nice playful touches." —Benjamin Brown, Harvard University

"It is, in short, written in the best scholarly tradition of the theoretically-sensitive case study, albeit written with much more grace than such studies." —Jorge Domínguez, Harvard University

"A very clearly written, enjoyable, and detailed account of what transpired." —Jeffrey Rubin, Tufts University

For Joan,
Gwen, and
Audrey

A Peace in Southern Africa: The Lancaster House Conference on Rhodesia, 1979

Jeffrey Davidow

Published under the auspices of the Center for International Affairs, Harvard University

Westview Press / Boulder and London

Westview Special Studies on Africa

Copyright © 1984 by Westview Press, Inc.

Published in 1984 in the United States of America by Westview Press, Inc., 5500 Central Avenue, Boulder, Colorado 80301. Frederick A. Praeger, President and Publisher

Library of Congress Cataloging in Publication Data
Davidow, Jeffrey.
    A peace in southern Africa.
    (Westview special studies on Africa)
    "Published under the auspices of the Center for International Affairs, Harvard University."
    Bibliography: p.
    Includes index.
    1. Lancaster House Conference on Rhodesia (1979)
2. Zimbabwe—Politics and government—1979–1980.
3. Zimbabwe—Constitutional history. I. Harvard University. Center for International Affairs. II. Title.
III. Series.
DT962.75.D38 1984   968.91'04   83-23393
ISBN 0-86531-703-8

Printed and bound in the United States of America

5   4   3   2   1

Contents

Preface

On June 7, 1979, President Carter issued a congressionally mandated report in which he stated that the conditions were not yet appropriate for the United States to lift economic sanctions against Rhodesia, which was technically still a colony in rebellion against Britain. The President's decision was not a popular one among many on Capitol Hill, where support for the Rhodesian regime of Bishop Abel Muzorewa, who had six weeks earlier replaced Ian Smith as Prime Minister, was increasing.

The Carter administration felt compelled to make several gestures to placate congressional opponents of its Rhodesia policy. The president himself would meet with Muzorewa on July 9 at Camp David, the first time an American chief executive had conferred personally with one of the contending Rhodesian nationalist leaders. Also Carter agreed to a congressional suggestion to send a U.S. diplomatic observer to Salisbury to monitor the progress of the Muzorewa regime. The United States had withdrawn all of its consular personnel from Rhodesia in 1970 in accordance with the sanctions program and had maintained no diplomatic or consular presence there since that time.

The gesture of sending an official was quickly pocketed and forgotten by the policy's critics; it would warrant little more than a footnote in even the most detailed tome exploring the Rhodesian drama. However, as the person chosen to go to Salisbury, I had a superb opportunity to watch history unfold. I was already familiar with the Rhodesian problem, having served as an officer in the U.S. embassy in South Africa from 1974 to 1976 and as the Department of State's Rhodesian desk officer for the following years. I had spent much of 1978–1979 as a congressional fellow of the American Political Science Association working in Congress, where I had seen some of the legislative debate on Rhodesia develop. Moreover, my middling rank was suitable to the task at hand. The dispatch of a more senior foreign service officer might have been distorted by several parties as connoting quasi recognition of the Muzorewa regime, an impression the State Department was anxious to avoid.

I arrived in Rhodesia in July 1979 and remained there, with only a few weeks outside of the country, for the next three years, a period that witnessed the end of a tragic civil war, the independence of Zimbabwe, and that state's first years as a majority-ruled nation.

Salisbury offered one of the most disadvantaged perspectives for trying to understand the negotiations that played themselves out at the Lancaster House Conference on Rhodesia of September through December 1979. Diplomatic communication was sporadic, the local news censored, and the reports filtering back from the various delegations so partisan as to make objective analysis impossible. The final Lancaster House accord ending the war and providing for a transition to independence came as something of a surprise to me, like most others resident in Rhodesia. I wondered how it had actually come about, but the press of events and work did not enable me to pursue the question seriously.

However, after leaving Zimbabwe, I was fortunate to spend the 1982–1983 academic year as a fellow of the Center for International Affairs at Harvard University. During that time I was able to take a closer look at the Lancaster House Conference. The more I studied it, the more impressed I was by the diplomatic victory that had been achieved in London. This book is a product of that process of discovery.

I am indebted to many individuals for their helpful comments and encouragement. Some are still closely associated with important Lancaster House actors, including the governments of Britain, the United States, and Zimbabwe. As it might unnecessarily complicate their lives to have to answer for the views expressed in this study, they shall remain thanked but unnamed at this time. I am very grateful for the support I received from the faculty and staff of the Center for International Affairs, headed by Samuel Huntington. Among the most patient and helpful collaborators there were Benjamin Brown, Jorge Domínguez, and Dov Ronen, director of the center's new and exciting African Research Program. Others who read early drafts of this work and offered helpful suggestions for which I am appreciative are Roger Fisher, Jeffrey Rubin, Robert Rotberg, George Betts, Bill Zartman, and Michael Schatzberg.

The absence of an abundance of footnotes sourcing specific bits of data about the actions or attitudes of the players at Lancaster House may trouble some readers of this volume. The only defense is that a good portion of the information presented was obtained during a decade-long process of immersion and osmosis and is not easily separable for purposes of scholarly attribution.

As a serving member of the U.S. Foreign Service, I have been obliged to obtain the clearance of the Department of State for the publication of these materials. Therefore, I must note, as I would in any event, that the views expressed, including those that might be wrongheaded or mistaken, are my own.

Jeffrey Davidow

1
Introduction

For more than three months in late 1979 British Foreign Secretary Peter Carrington chaired a conference at Lancaster House in London that the British government had convened to find, once and for all, a solution to "the Rhodesian problem." At its outset and through most of its course few observers or participants gave the conference much chance of success. Most doubted that a settlement acceptable to all of the parties gathered there could be devised that would bring an end to the war in Rhodesia and guide the country from minority white domination to majority African rule.

The pessimism was understandable. For fifteen years successive British governments had failed in efforts to convince Ian Smith, the then Prime Minister of Rhodesia, to relinquish the control that he and approximately 200,000 white settlers maintained over Rhodesia's government and its black population, which by 1979 numbered approximtely 7 million. Rhodesia had seemingly inscribed itself on the permanent agenda of the world's political-ethnic conundrums—Northern Ireland, Cyprus, the Middle East, South Africa—unamenable to human persuasion or reason.

Nevertheless, the conference did succeed. On December 21, the head of the Salisbury delegation, Bishop Abel Muzorewa, the leaders of the Patriotic Front, Joshua Nkomo and Robert Mugabe, and Lord Carrington signed an agreement that contained a constitution for the independent state of Zimbabwe, ceasefire provisions to end the war, and transitional arrangements to guide the country through a brief period of British interim administration. Why did the Lancaster House Conference succeed? The answer is twofold: Rhodesia was ready for a settlement; and effective British diplomacy converted the favorable situation into the reality of an agreement acceptable to all of the parties.

The road to Lancaster House was a rocky one. British elections in May 1979 brought a Conservative government headed by Margaret Thatcher to power. The Tories had already signaled their intention to recognize the Rhodesian regime of Bishop Muzorewa that had come to power as a result of an "internal settlement" between Smith and some of the country's black domestic politicians. The political leadership of the Patriotic Front had not participated in the internal settlement, nor in the April 1979 elections that gave an overwhelming parliamentary majority to the bishop. World opinion heavily favored the Patriotic Front's rejectionist stance, but Mrs. Thatcher was scornful of the "Marxist terrorists," led by Nkomo and Mugabe, and seemed destined to lift the economic sanctions imposed years earlier on Rhodesia and to move toward a formal granting of independence.

But the spring and early summer of 1979 witnessed the education of Margaret Thatcher. Her principal tutor was her urbane and skilled foreign secretary. By the time of the Commonwealth Heads of Government meeting in Lusaka in August 1979, she had reluctantly accepted that recognition of the Muzorewa regime would be a major political blunder, isolating Britain on this issue from her allies, notably the Carter administration, and further alienating Commonwealth and Third World opinion. Mrs. Thatcher's surprise diplomacy at Lusaka began a new chapter in Britain's search for a settlement.

The delegations that assembled on September 10 at Lancaster House in London carried with them different objectives. For the Salisbury team, led by Bishop Muzorewa and including Ian Smith, the principal goals were to obtain the lifting of sanctions and rapid recognition. Equally important to Muzorewa was his firm desire not to give up any of the power, limited though it may have been, that his premiership conveyed to him. Nkomo and Mugabe came to London intent on maintaining the negotiating stance they had held firmly since the failed Geneva Conference of 1976. The Patriotic Front argued that power must be transferred to it at the outset of any transitional period to guarantee that elections for the country's independence government be conducted in an atmosphere unthreatening to its political and guerrilla forces.

The two delegations had limited characteristics in common. They profoundly distrusted each other. Neither believed that Lancaster House would result in a settlement acceptable to all parties, but neither was willing to abdicate the field of negotiation to the other. Both recognized that the Rhodesian war had reached a stalemate in which rapid victory was impossible, and each believed that it enjoyed majority support among their country's African population. Compelling the two delegations to seek a settlement were important external actors or forces: for the Muzorewa regime the most important exogenous element was the parlous state of the country's economy, drained by the war effort and buffeted by world recession, high prices for imported oil, and the effects of international economic sanctions; the principal compelling force operating on Nkomo and Mugabe was the combined weight of the Front Line states, which had tired of the war and saw in the Lancaster House Conference a plausible and honorable way out for themselves and the Patriotic Front. Both delegations were seriously divided: the Salisbury team was a hybrid conglomeration of often antagonistic whites and blacks; the Patriotic Front was little more than a negotiating vehicle for two political-military movements whose long relationship had been marked by hostility and tension.

The British team on the other hand was well unified, vertically integrated from Mrs. Thatcher down to the small group of Foreign and Comonwealth Office (FCO) professionals who assisted Carrington. The foreign secretary himself was a skilled and experienced negotiator and his team had devised, even before the conference, a tactical plan. The major elements of the British plan were emphasis on U.K. centrality in the negotiating process; almost dictatorial conference management;

and a step-by-step approach. The latter was designed to divide the issue into more easily negotiable elements, and was closely related to another major part of the British approach, a willingness to accept a "second-class solution."

That term served as shorthand for a separate British deal with the Muzorewa delegation. Carrington preferred an all-parties agreement, but doubted the prospects of arriving at one. Nevertheless, he was willing to try. In this effort his perceived willingness to accept the second-class solution was his greatest tactical weapon, for he was able to keep the Muzorewa delegation on the negotiating hook by continuing to threaten the Patriotic Front with a separate settlement with the bishop. The Patriotic Front's leaders responded to the threats, much encouraged to do so by the Front Line states.

Carrington's negotiating position was markedly strengthened by the situational factors surrounding the conference: its London venue and the wide panoply of tools—intelligence gathering, press manipulation, tacit bargaining played out in Parliament—that he put to use. However, the situational factors would not have been significant without Carrington's impressive negotiating skill, particularly evidenced in the manipulation of threats and promises revolving around the second-class option. The Muzorewa and Patriotic Front delegations were outgunned and outmaneuvered by the British in what must be regarded as one of the United Kingdom's few recent diplomatic triumphs.

This study will seek to explain what happened at Lancaster House and to offer some thoughts on how that historical event might be used by scholars and theoreticians to develop new insights on a special kind of negotiation that can be called dominant third-party mediation. Chapter 2 sets the stage for the conference. Chapter 3 outlines the composition of the delegations, their strengths, weaknesses, and objectives. The next three chapters deal with the successive negotiating phases of the conference mandated by Britain's insistence on a step-by-step approach: the constitution; the transitional arrangements; and the ceasefire. Chapter 7 outlines how the settlement actually worked when placed into operation and is followed by a discussion in Chapter 8 that isolates and identifies those factors accounting for Carrington's success. Finally, Chapter 9 offers some thoughts on the nature of negotiation and the characteristics of dominant third-party mediation. Briefly stated, such mediation, although not common, is not unique to the Lancaster House Conference. It is a possible course of mediation when external circumstances conducive to the resolution of a conflict are used to advantage by a third party, whose authority to intervene is accepted by the disputants, and who is sufficiently interested, powerful, and skillful to manipulate the other actors into accepting his suggestions for resolution.

In retrospect, Lancaster House appears to have been an easier victory for good sense than in reality it was. What is absent from the historical perspective is the fog of uncertainty, confusion, and misperception that hung over the conference, preventing most there from obtaining a clear view of what was actually happening.

Lancaster House had many of the ingredients of high theater: colorful personalities, cliff-hanging plots, intrigue, humor, violence lurking in the wings, and more—which will become apparent in what follows.

2
Setting the Stage

By the time Margaret Thatcher became Britain's Prime Minister in May 1979, the Rhodesian problem had been a standard fixture of British political life for a decade and a half. Ian Smith, the Prime Minister of that self-governing territory, had issued its unilateral declaration of independence (UDI) from Britain in 1965, after both Labour and Conservative governments had made it clear that formal independence could not be granted to Rhodesia without a firm commitment by a Rhodesian government to move the country toward majority rule. Intent on maintaining unfettered white minority control, Smith rejected British blandishments and issued his declaration of independence, its language made purposefully reminiscent of the American document of 1776.

Successive British governments struggled unsuccessfully to negotiate a solution to the problem. Settlement had been nearest at hand in 1971 when Smith and Conservative Foreign Secretary Sir Alec Douglas-Home agreed to proposals that posited a slow increase in the number of African voters over an undefined length of time until blacks, who comprise 97 percent of the population, would be able to obtain a slight majority in the Rhodesian parliament. Britain insisted, however, that implementation of the Home-Smith deal be contingent upon a test of African public opinion to determine its acceptability. For this purpose a commission led by Lord Pearce visited Rhodesia in the first months of 1972. Rhodesia's contending nationalist parties temporarily buried their hatchets and selected Methodist Bishop Abel Muzorewa to present the views of politicized Africans to Pearce under the umbrella of the newly created African National Council. Muzorewa's presentation and their own soundings compelled the commission to conclude that although many Africans might not comprehend the agreement's fine points, they categorically rejected the concept of Britain's granting independence to a government they profoundly distrusted. NIBMAR, No Independence Before Majority Rule, became the African political motto. Much to the bitter disappointment of many in the Rhodesian government and the Conservative party, the agreement was shelved. Britain adopted a relatively low posture in relation to the Rhodesian question for the following four years.

What brought Rhodesia back into the world spotlight and onto the British agenda once again were the major geopolitical changes occasioned by the dissolution of Portugal's African empire. By 1975 a Marxist African government had established itself in an independent Mozambique, thus opening a new front on Rhodesia's eastern border for the guerrilla forces that had been mounting an increasingly intense, but still low-scale, insurgency from bases in Zambia since late 1972. A civil war among the three contending heirs to Portuguese power in Angola had drawn in the superpowers, with 15,000 to 20,000 Russian-supported Cuban troops aiding one faction and the United States covertly arming the other two. In late 1975 the U.S. Congress forced President Ford and Secretary of State Henry Kissinger to desist from covert activity in Angola. Kissinger determined that further Soviet-Cuban expansionism in the area could best be prevented by eliminating potential fields of battle. His gaze naturally fell on Rhodesia, where Russian and Chinese arms were already being used by the guerrillas.

Kissinger's advent on the scene introduced the United States as a major actor in the effort to bring about a Rhodesian settlement. The United States' most important prior contribution to the Rhodesian problem had been the passage in 1971 of the Byrd Amendment, introduced by Senator Harry Byrd of Virginia, which permitted U.S. imports of chrome and other Rhodesian minerals in open contravention of the

United Nations' sanction resolutions, for which the United States had voted. The Byrd Amendment strengthened the Smith regime and, at the least, did not promote a Rhodesian settlement. During a visit to southern Africa in September 1976, Kissinger met with Ian Smith and obtained from him, through the artful use of personality, ambiguity, and pressure (largely South African in origin), his agreement to attend a British-chaired conference that would work out the details of a transitional plan to bring Rhodesia to majority rule within two years. The "Kissinger Plan," actually more a British than a U.S. product, was seriously flawed in many respects. The resulting Geneva Conference of October–December 1976 brought together the same actors who would assemble three years later at Lancaster House. Poorly prepared, the conference mired hopelessly in peripheral debate, failed to reach minimal agreement, and ended in total failure. The scar of Geneva, the British impotence displayed there, was to weigh heavily on the United Kingdom in future negotiations. The desire to avoid another Geneva was an absolute imperative for Lord Carrington at Lancaster House.

At the beginning of 1977 the Geneva Conference's failure seemed to forebode another period of reduced British diplomatic activity. However, fate intervened: Foreign Secretary Anthony Crosland died and was replaced by a young, bright, and ambitious Labour MP from Plymouth, David Owen. He found in the new U.S. administration of Jimmy Carter a high-level interest and commitment to an aggressive Rhodesian diplomatic effort. In Salisbury on September 1, 1977, Owen and U.S. Ambassador to the United Nations Andrew Young revealed new Anglo-American proposals for a settlement. The plan included principles for an independence constitution and a transitional period under the stewardship of a British resident commissioner who, with the assistance of a United Nations peacekeeping force, would organize one-man, one-vote elections leading to independence before the end of 1978.

At the insistence of President Julius Nyerere of Tanzania, who had visited Washington in July, the plan also called for the creation, during the transitional period, of a new Zimbabwe army to be drawn predominantly from the guerrilla forces. Owen believed that this provision would spell doom for the proposals, but President Carter, who had given Nyerere his personal word on the issue, was adamant. The language was adopted. Owen was right. The military provision and those relating to involvement of the much-despised United Nations were only two of the elements that were particularly objectionable to the Smith regime. For the Patriotic Front, led by the exiled Joshua Nkomo of the Zimbabwe African People's Union (ZAPU) and Robert Mugabe of the Zimbabwe African National Union (ZANU), the plan was equally unacceptable. They specifically objected to numerous elements, but most vociferously to those that would give Smith and his officials a continuing, and seemingly too large, share of power during the transition. However, as neither the Patriotic Front nor the Smith regime wanted to be portrayed as the most recalcitrant element in the Rhodesian imbroglio, neither rejected the plan outright. With no better suggestions at hand, the United States and the United Kingdom kept pushing the proposals without success for the next eighteen months. That diplomatic effort did have some value in that it kept a negotiated settlement on the table as a topic for intermittent discussion and as an ultimate goal. Further, like the Geneva Conference before it, the Anglo-American failure was to be mined for lessons by the Conservative government after its assumption of power in May 1979.

Though unsuccessful, the 1976–1979 diplomatic efforts had helped change the Rhodesian scene dramatically. Smith, in accepting the Kissinger plan, had at least acknowledged the possibility of majority rule within a few years. The growing intensity of the guerrilla war, domestic economic difficulties, and a creeping realization that Rhodesia could not obtain international legitimacy without a preponderantly African government, propelled Smith into capitalizing on divisions within the African nationalist movement. The result was the internal settlement of March 1978, of which the principal black participant was Bishop Muzorewa, no longer the apolitical cleric but an ambitious claimant for the same position of national leadership to which both Nkomo and Mugabe, as well as many others, aspired.

The "external" leaders of the Patriotic Front were tendered pro forma invitations to participate in the internal settlement, if they would lay down their arms, that is, surrender. Their rejection was foreordained, although Smith continued to harbor hopes that he could split Nkomo from Mugabe. To this end he met secretly with the ZAPU leader, but without success. Smith saw in the internal settlement a path to respectability. A black government would compel the United Nations, or at least leading Western nations, to end Rhodesia's international isolation and lift the economic sanctions that had been made more onerous by the repeal in March 1977 of the Byrd Amendment. The internal settlement, Smith argued, would provide the new government with the political legitimacy necessary to convince some guerrillas and their international supporters of the futility of continued fighting. The termination of sanctions would provide the government with the economic wherewithal to prosecute the war more vigorously against those guerrillas who still held out. It would also help restore health to the seriously damaged Rhodesian economy. The sanctions themselves were not debilitating, but their cumulative effect when added to the impact of the world recession, rising oil prices, and the growing costs of fighting the war placed Rhodesia's economy in a parlous state, heavily dependent upon the financial largesse of South Africa.[1] As the country's real gross domestic product declined after 1974, its defense spending rose rapidly, forcing the regime to go increasingly into debt. Without access to Western loans and markets that recognition and the lifting of sanctions would bring, further decline was inevitable.[2]

The March 1978 agreement provided for a government by committee for the rest of the year in which black and white co-ministers would share portfolios. Smith would remain as Prime Minister. At the end of the period, elections would be held for a parliament, to be composed of seventy-two black and twenty-eight white members, which, in Westminster fashion, would choose a government to rule under an agreed-upon constitution. That document would be drafted in the months preceding the election. In fact, it took somewhat longer than expected to negotiate the details of the constitution. It rapidly became apparent that Smith and his Rhodesian Front party were intent on maintaining political control in the hands of the white minority.

After much acrimonious debate the new constitution was promulgated in January 1979. Its provisions revealed the degree to which Muzorewa and the other black participants in the internal settlement had given in to Smith's demands. The document stipulated requirements for high office in the police, military, judiciary, and civil service that virtually guaranteed that few Africans would be able to assume positions of responsibility in the foreseeable future. White-dominated commissions were given major responsibilities for supervision of the four areas, limiting the control of responsible African ministers. Their twenty-eight (out of 100) reserved seats guaranteed the whites a large role in parliamentary affairs and, through a proportional system of portfolio allocation, important cabinet positions. In all, the constitution contained over 120 "entrenched," that is unamendable, provisions designed to protect the position, privilege, and power of whites. The African majority was even denied the satisfaction of giving the government created by the new constitution the long-sought-after name of Zimbabwe. In a move redolent of the desire to preserve the past, Smith first insisted on no name change, but then grudgingly accepted the hybrid and, for many, insulting appellation of Zimbabwe-Rhodesia.[3] The constitution's weaknesses reflected the political facts of life. Muzorewa was dependent on the white-controlled civil service and military to maintain himself in power and to prosecute the war against the guerrilla forces of Nkomo and Mugabe.

Despite the obvious deficiencies of the constitution, the internal settlement had attracted strong support in the British Conservative party, not only from right-wing members who had traditionally lionized Ian Smith, but also from more moderate elements who welcomed the prospect of a government led by an American-trained Methodist bishop as an appealing

alternative to the Marxist regime seemingly promised by the
Patriotic Front "terrorists." The Conservative party's 1979
election manifesto promised that

> if the six principles, which all British Governments have sup-
> ported for the last fifteen years, are fully satisfied following the
> present Rhodesian elections, the next government will have the
> duty to return Rhodesia to a state of legality, move to lift
> sanctions, and do its utmost to ensure that the new independent
> state gains international recognition.[4]

To set the stage for the return to legality, the Conservatives
dispatched a commission under the chairmanship of Lord
Boyd, a former colonial secretary, to observe the Rhodesian
election of April 17–21, 1979.

The election was an impressive success for the Smith-
Muzorewa alliance. Nkomo and Mugabe had refused the government's call to lay down their arms and take part in the
campaign. They pointed to the practical impossibility of participating in an electoral process run by their sworn enemies.
Total mobilization of the country's white police and military
reserves and the established security forces effectively prevented the guerrillas from carrying out their threat to disrupt
the polling. Given their first opportunity to vote, the African
population responded with considerable enthusiasm. In five
days 1.75 million Africans cast ballots. Muzorewa's party was
the big winner, capturing fifty-one of the seventy-two African
seats. Smith's Rhodesian Front had easily won all twenty-eight
white seats in elections held a few days earlier.

The report of the Boyd Commission was released on May
24 after British general elections had brought Mrs. Thatcher
to power. The report expressed some reservations, but, nevertheless, pronounced that "neither censorship nor martial law
nor the refusal of the Patriotic Front to participate in the
election invalidated the results. . . . Though the election may
not have been fully free because of war conditions it was at
least as fair and free as possible. Moreover, it could be taken
as 'a kind of referendum' on the constitution."[5] Obliged by

her own party's manifesto, and her own prejudices buttressed by the Boyd report, the new British prime minister's next steps—lifting of sanctions and return to legality through recognition of the new government—seemed clear. But Margaret Thatcher did not take those steps for reasons that were both many and compelling.

At base, Mrs. Thatcher realized her instincts had to be tempered with caution. For although the Muzorewa victory was popular among British Tories, that sentiment was not shared by the rest of the world. Indeed, international reaction had been almost universally hostile. Resolutions in the United Nations and other multinational fora had condemned the internal settlement and then the elections as malevolent farces. In addition to predictable Third World opposition to any steps to legitimize the Muzorewa regime, the message was passed from Washington to the new British administration that, despite congressional pressure, President Carter was not prepared to lift sanctions or to otherwise assist Muzorewa, even if Britain were to do so. The probable lack of any measurable international support weighed heavily on Mrs. Thatcher: leading is no fun if nobody will follow. Another major consideration was the August 1979 Commonwealth Heads of Government Conference, scheduled to be chaired in Lusaka, Zambia, by President Kenneth Kaunda, a firm supporter of the Patriotic Front. Any British action with respect to the new Muzorewa regime, therefore, had to be weighed carefully with respect to the larger international environment and with specific reference to the Commonwealth, which was overwhelmingly anti–internal settlement.

An understanding of the need for caution and prudence had developed among leading Tory politicians even before the Conservatives were voted into power. Some saw the dangers inherent in an immediate recognition of the Muzorewa government. One who did not, but later had the intellectual grace to revise his opinion, was Miles Hudson, a member of the Boyd Commission and since 1965 the chief analyst of Rhodesian affairs in the Conservatives' research department. He has since written:

On balance, and he certainly did not think so at the time, it
is the author's opinion that the government was probably right
in the circumstances to do what it did, in spite of its Manifesto
commitment. If the Rhodesian Government had been recognized
and sanctions had been dropped, the war probably would not
have ceased. Britain would not have been able to convene, let
alone bring to a successful conclusion, a new conference whether
at Lancaster House or anywhere else. In the end, after even
more bloodshed than actually occurred, Mugabe would probably
have come to power with a great grudge against Britain, whose
international position would have been much undermined.[6]

Hudson noted as early as April 1979, that is, before both
the Rhodesian and British elections, that leading Conservative
figures, including Peter Carrington and Francis Pym, the two
front-runners for the post of foreign secretary, were hedging
their public statements on Rhodesia in order to provide themselves with more running room. It was Carrington who got
the nod from Mrs. Thatcher to be her foreign secretary. (Pym,
who was given the defense portfolio, would succeed him upon
his resignation over the Falkland Islands' crisis and then be
dropped from the cabinet himself following Mrs. Thatcher's
landslide reelection in June 1983.)

A man of considerable governmental experience, Carrington
knew more about international topics than most of the Tory
leadership. His knowledge of Africa, and particularly Britain's
post-colonial position there, both political and economic, was
highly developed. His predecessor, David Owen, credited Carrington with being "the only senior Tory who understood
Africa. He always understood what I was trying to do, and,
within the rules of the game, always played with honor—
unlike many other Tories, who didn't and who tried to exploit
the situation. I always hoped he would be my successor."[7]

Once in office Carrington quickly came to agree with his
Foreign and Commonwealth Office professionals that Britain
must rapidly confront the Rhodesian problem. A new initiative
was called for, as Carrington explained several years later to
a reporter:

You couldn't leave things as they were. You had got to the end of the road with the Anglo-American proposals, and recognition of the Bishop's regime would have led to the most appalling problems—not least the isolation of Britain by the rest of the world, including the United States. And it would have intensified the war in Rhodesia. There would have been much more bloodshed. It would have settled nothing, and, incidentally, would have brought the sorts of things that everybody has been trying to avoid in Southern Africa that much nearer—the Soviet involvement, the East German involvement, and all the rest of it.[8]

At a press conference on May 14, the new foreign secretary signaled the climb-down from the Conservative party's manifesto by announcing that his government's Rhodesia policy would be formulated only after extensive talks with Commonwealth partners, African and European statesmen, and the United States government.[9]

A week later he had an opportunity to find out first-hand what the Carter administration had in mind when Secretary of State Cyrus Vance visited London. The two men had known each other for almost two decades, first working together when Vance was John Kennedy's Secretary of the Army and Carrington was Harold Macmillan's First Lord of the Admiralty. Their May 21 meeting was a tour d'horizon of world trouble spots, which naturally included Rhodesia.[10] Accompanying Vance were Dick Moose, the Assistant Secretary of State for Africa, and Anthony Lake, the Department of State's Director of Policy Planning, two of the principal architects of the Carter African policy. Both men were concerned that Mrs. Thatcher might move swiftly to lift sanctions against Rhodesia as a prelude to recognition of the Muzorewa regime. Carrington reassured them that precipitous action was not in the cards, but he left no doubt about Mrs. Thatcher's inclinations and flatly stated that she would not be prepared to urge the continuation of sanctions when Parliament reviewed them in November.

Vance briefed Carrington on the growing conservative mood in Washington, particularly in Congress. It is unlikely, though possible, that he would have confided that the desire to be more accepting of the Muzorewa regime was also shared by the president's National Security Advisor, Zbigniew Brzezinski.[11] He did, however, note that Carter would consider lifting sanctions if the Salisbury regime took specific steps to overcome the deficiencies of the internal settlement. He enumerated these as constitutional revision, progress toward an all-parties conference without giving the Patriotic Front a veto over whether such a conference should occur, and new, internationally supervised elections.

Carrington was dismayed by the specificity of the U.S. requirements and argued for a more general conditionality that would give him more flexibility to maneuver. Vance may have shared the concern of Moose and Lake that the United States not allow Britain too much leeway lest it be taken advantage of by Mrs. Thatcher to reach an unsatisfactory agreement with Muzorewa. Nevertheless, on his return to Washington Vance presented the president with the option of general versus specific conditionality, and Carter chose the former. The decision angered some liberal members of the House who were the bulwark against congressional sentiment in favor of lifting sanctions and were distrustful of British motives. Further, by not publicly declaring what specific steps the U.S. government wanted the Muzorewa regime to take, Carter made defense against arguments that indeed Muzorewa was moving toward complete majority rule that much more difficult. That, however, was a problem for the United States. For Carrington, his meeting with Vance was a success in that he was given the running room he needed, free from an overly restrictive U.S. guiding hand.

Carrington went to work immediately. He dispatched Sir Antony Duff, Permanent Deputy Undersecretary of the Foreign and Commonwealth Office, to Salisbury to talk to Muzorewa. Shortly thereafter it was announced that Lord Harlech, a distinguished British diplomat who had served as ambassador to Washington and had been deputy chief of the Pearce Commission, would tour several African states to solicit views on the Rhodesian question.

The Harlech trip was a key element in what British officials in the spring and early summer of 1979 privately termed "the education of Maggie Thatcher." For though Carrington maintained his strong inclination to support Muzorewa, he had quickly comprehended the potential gravity of recognizing the bishop's regime and had carefully sought to bring the fiercely "antiterrorist" Mrs. Thatcher around to a more balanced view. Ironically, Carrington and the FCO professionals used the same incremental approach to this educational process that they were to utilize at the Lancaster House Conference. Mrs. Thatcher was led, step by step, to the formulation of a new Rhodesian policy that was to crystallize at the Lusaka Commonwealth meeting. The new Prime Minister's faith in her foreign secretary, whom she then viewed as much more versed in international affairs than herself, was considerable. She therefore accepted his suggestion to send Harlech to Africa, a trip that she must have realized was unlikely to result in a ringing endorsement of the Tory manifesto.

Indeed, on his return in June, Harlech frankly noted the opposition of African leaders to the Muzorewa government and to any British inclination to recognize it. But Harlech also found, and purposefully emphasized in his report to Mrs. Thatcher, that many African leaders had acknowledged that the new Rhodesian political configuration could be seen as a form of progress if only because Ian Smith was no longer premier. If the more pernicious elements of the constitution were eliminated and a way devised for the Patriotic Front to participate, the Rhodesian crisis might possibly be resolved. What was necessary, stressed the Africans and Harlech, was an aggressive British diplomatic effort, an exercise of its legitimate authority to wrap the problem up.

The idea of a more "manly" approach appealed to Mrs. Thatcher, in her desire to return Britain to the ranks of world power. With the active support of Commonwealth Secretary General Sir Sridath Ramphal, Carrington and the FCO professionals were well on the way toward a new negotiating drive before Mrs. Thatcher and Carrington accompanied the Queen to Zambia in early August. The welcome that President Kaunda accorded Mrs. Thatcher was a cool one. But if Kaunda was unaware or suspicious of Mrs. Thatcher's willingness to embark on a new diplomatic initiative to avoid the isolation that recognition of Muzorewa would engender, others, notably President Nyerere of Tanzania, were not. However, before progress could come, there awaited a potential disaster.

On the Commonwealth conference's opening day Nigeria nationalized British Petroleum's oil production and marketing facilities in that country, ostensibly in relation to BP's continued supply of oil to South Africa. However, in Lusaka, Nigerian Minister of Foreign Affairs Henry Adefope acknowledged to the press that the move was designed to make "Britain look at the Zimbabwe problem in a more realistic way." Carrington responded angrily, reportedly telling Adefope at the conference's opening day garden party that "you'll regret the timing of this."[12] Carrington's ire was undoubtedly intensified by his concern that the Nigerian action would provide just the provocation necessary for Mrs. Thatcher to jump off the negotiating bandwagon that he and Ramphal were beginning to push. In fact, she did not bolt. In some measure her decision was clearly influenced by her desire not to embarrass the Queen, who had let Mrs. Thatcher know that she would not be pleased by disarray or vituperation in Lusaka. Moreover, in her first sustained contact with the Commonwealth's African leaders she found them to be more reasonable, less tub-thumping and doctrinaire, than she had imagined. Finally, the presence of Australian Prime Minister Malcolm Fraser, an ideological conservative but virulently anti–internal settlement, provided a steadying influence on her.

After initial moderate and compromising speeches by Mrs. Thatcher and President Nyerere of Tanzania, a small working group was formed that included those two, Adefope, Kaunda, Fraser, Prime Minister Manley of Jamaica, and Ramphal. Carrington, of course, sat in as well. When, following a series of substantive and cordial meetings, Ramphal produced principles for a Rhodesian settlement, cleared in advance by Carrington, who thought that Britain's purposes might be better served by having the ideas presented as a Commonwealth draft, agreement was forthcoming. Many chiefs of state had come to Lusaka expecting a battle and were caught off guard by the now educated Mrs. Thatcher's surprise diplomacy.[13] The conference unanimously accepted the working group's draft. In the most important elements of the nine-point plan the Commonwealth

> Fully accepted that it is the constitutional responsibility of the British Government to grant legal independence to Zimbabwe on the basis of majority rule;
>
> Accepted that independence on the basis of majority rule requires the adoption of a democratic constitution including appropriate safeguards for minorities;
>
> Acknowledged that the Government formed under such an independence constitution must be chosen through free and fair elections, properly supervised under British Government authority, with Commonwealth observers; [and]
>
> Welcomed the British Government's indication that an appropriate procedure for advancing towards these objectives would be for them to call a constitutional conference to which all the parties would be invited.[14]

Within a week Carrington issued invitations to a constitutional conference, to be held at Lancaster House in London, to Bishop Muzorewa and the leaders of the Patriotic Front. Only a little more than three months had passed since the Conservative party, pledged by its own campaign platform to recognize the Muzorewa regime, had taken power. Now a new and major diplomatic effort to bring about a negotiated settlement to the Rhodesian problem was under way.

3
The Delegations

Who came to the Lancaster House Conference and why? What were the compositions of the delegations and their respective objectives? What strengths and weaknesses did the delegations bring with them?

The British

The major British objective, simply put, was to remove the Rhodesian problem as a constant thorn in the United Kingdom's side. U.K. representatives at international gatherings had wearied of the continuing and often bitter denunciations aimed at their country over its alleged abdication of responsibility for its colony. A negotiated settlement would open the way for better relations with black Africa and the Third World. In addition, Rhodesia would cease to be a highly conflictive issue within the Conservative party, one that diminished the necessary Tory unity for dealing with other domestic and foreign problems.

The small team that Carrington put together to assist him at Lancaster House was notable for its ability. Ian Gilmour, the Deputy Foreign Secretary, was the logical choice as the conference's deputy chairman as he would be obliged to defend and explain the proceedings in the House of Commons, where Carrington, as a member of the House of Lords, could not speak. It had been taken as an early sign of Mrs. Thatcher's respect for Carrington that she had acquiesced in his request that Gilmour be given the number two position at the Foreign and Commonwealth Office, for she did not particularly like the droll Gilmour as a person and found his politics too "wet," that is, too soft or too liberal. The senior professional diplomat on the conference scene, Sir Antony Duff, had demonstrated through his long career abilities as a superb tactician with an easy manner. Robin Renwick, the head of the Foreign Office's Rhodesia department, combined a hard-driving nature with brilliance and detailed knowledge of the Rhodesian scene. Other Whitehall professionals rounded out the U.K. team. They established with Carrington a sense of mutual confidence that flowed from the traditional Tory predilection for placing considerable confidence and responsibility on Britain's career diplomats.[1]

Guiding them was Carrington himself. Well versed in international affairs and negotiation, his personality attracted much press attention. Journalists who described him seemingly chose from a list of adjectives that included charming, aristocratic, urbane, humorous, occasionally self-deprecating, incisive, quick to anger (though not averse to using mock anger as a negotiating tool), self-assured, and authoritative. Carrington met with his team every morning to chart the day's tactics. Although he delegated tasks, notably to Duff and Renwick, he maintained a tight rein. In these meetings he encouraged open exchanges of views and particularly welcomed the interventions of Gilmour, who played the self-appointed role of witty devil's advocate.

Carrington also met every day with Mrs. Thatcher. These chats were known to FCO professionals as the foreign secretary's "other negotiations." For although the prime minister had been educated, she nevertheless remained deeply hostile to Nkomo and Mugabe and a partisan of "the game little bishop." Carrington's task, therefore, was to keep her not only informed, but also supportive of his positions during the conference and to do so in a manner that maintained his credibility. He succeeded. During the conference there was no press speculation about differences between him and the prime minister. Vertically integrated from Mrs. Thatcher down to the FCO professionals, the British team contrasted markedly with the other two delegations.

Nevertheless, Carrington's position was not unassailable. He was to come under attack on several occasions by the parliamentary opposition, but these assaults were easily parried. Of more pressing concern to him, and some observers believe of too great a consideration given his own strength and position, was the reaction of his own party's right wing. No issue so excited Tory back-benchers as did Rhodesia. Ian Smith had consistently maintained the support of the party's far right, led on this issue by MP Julian Amery. And after the internal settlement agreement of 1978, the ranks of Conservatives eager to lift sanctions swelled appreciably. In November 1978, 116 Tories in the House of Commons broke ranks with the party leadership, the largest number ever to do so, and voted against renewing the legislation necessary to continue sanctions for another year. As party leader, Mrs. Thatcher, who had urged the loyal opposition to abstain, was obliged to remove two front-benchers from their party positions as an act of demonstrable discipline, even though her own views on Rhodesia were similar to those of the dissidents.

Carrington had been a Conservative all of his adult life, serving in Tory governments since 1951. But although *in* the party, he was really not *of* it, especially in its latest, most conservative form under Mrs. Thatcher. The foreign secretary purposely isolated himself from governmental debates on social and economic policies so as not to have to confront his prime minister. He generally felt uncomfortable with his own party's right wing, precisely the element that was most vocal about Rhodesia. One perceptive journalist has written that

> He probably distrusts the right more than the left. He is rather patronizing about the left, which he thinks is self-evidently wrong about most things, but he is distinctly hostile to the Tory right—to both its elitist and its so-called suburban strands. Even when he was the chairman of the Party—during the government of Edward Heath, in the early nineteen seventies—he retained 'a certain contempt for it,' according to one of his political

confederates. He was often heard to refer to the Tories as 'they.' Mrs. Thatcher's political universe is no closer to his than is Labour's working-class electorate.[2]

Given this background, it is not surprising that with a Tory Congress scheduled for early October, the possibility of a revolt within the ranks weighed heavily on Carrington's thinking and management of the Lancaster House Conference. As events developed, the foreign secretary was able to manipulate the Tory right wing into impotence or unwitting assistance, but he could not foresee this at the outset.

The British did not develop a blueprint prior to the conference that would dictate each move in advance. In fact, as Lancaster House progressed, they found themselves encountering unplanned-for situations, making commitments not entertained, or even thought feasible, at the outset. Nevertheless, there were elements of Carrington's conference tactics that were decided upon before the two other delegations arrived in London and that were to be critical to the conference's ultimate success. They were: (a) emphasis on British centrality in the negotiating process, which mandated the relegation of other governments to the periphery; (b) strong, almost dictatorial, conference management; and (c) a step-by-step approach.

The recent negotiating past heavily influenced British thinking. The Tories viewed the Anglo-American diplomatic effort of 1976–1979 as inherently unwieldy. Getting one government, with its many disparate bureaucracies, political forces, goals, and policies to march in the kind of lockstep successful negotiating demands is difficult enough. The Tories recognized that getting two governments to march to the same drumbeat is infinitely more difficult. Moreover, U.S. "muscle," the presumed ability to pressure South Africa to use its influence on Smith, which had so intrigued David Owen, was far less attractive to the Conservatives, who viewed the Pretoria government as at best, a potential conference ally and at worst, a potential settlement spoiler that had to be placated and not enraged by implicit or explicit threats. Further, sharing the negotiating load with any other country was seen by the Tories as an embarrassing and unwarranted derogation of British authority. Finally, Carrington himself had had unhappy experiences with U.S. negotiators when, as minister of defense in 1971–1972, he had been in charge of negotiating a new British military base agreement with the slippery Maltese government of Dom Mintoff. Repeatedly, his firm British positions had been undercut by less resolute U.S. interlopers.[3]

Whatever other motivations may have existed, the British were intent on demonstrating that the conference would be their show from start to finish. But they realized that a brusque dismissal of nonconference participants would be both impossible and unwise. Though relegated to the periphery, the active support for, or at least the acquiescence in, a settlement by the United States, the Commonwealth Secretariat, the Front Line states, and South Africa was imperative. The management of Britain's relations with these out-of-Lancaster House actors was to consume a great deal of Carrington's time, energy, and patience. In the end, it is doubtful that success could have been achieved without the intervention at crucial junctions of some of the supporting cast of onlookers and kibitzers. In this regard, the Commonwealth Secretariat and the Mozambican and U.S. governments would prove to be key players.

Carrington felt that the British had lost control at Geneva, contributing to that conference's failure. Lancaster House's proceedings would remain firmly in British hands, with Carrington acting autocratically if necessary. Procedural objections would be dealt with summarily and British drafts would serve as the basis for discussions. In practice, Carrington and his team developed a negotiating style that ameliorated the tough exterior with a more flexible approach in camera, but Carrington was to be adamant throughout the conference in preserving the prerogatives of the chair.

A step-by-step approach was seen to have multiple advantages. By dealing with the less contentious issues first, momentum could be generated, which would help in confronting thornier topics. More importantly, partial agreements would serve as a hook to keep the parties negotiating. In case of failure part way through the conference, partial agreements could provide the basis for Britain's next steps. Robin Renwick, a key member of the Carrington team, has explained the choice of the constitution as the first issue to be discussed in the following manner:

> The attainment of majority rule was what the war was supposed to be about. If it were possible to achieve an independence constitution which, indisputably, provided for that, the conflict would be reduced to the dimensions of a struggle for power; and the pressure would be greater on the parties for that competition to be settled by other means. The momentum thus achieved towards a settlement could help the negotiations through the later and still more difficult phases. . . . It was no less essential to the chances of success to insist that the Conference should be progressive—that it should not move on to the second stage (the interim arrangements) until it had settled the question of the independence constitution. For otherwise it was liable to degenerate, as earlier discussions had done, into an endless debate.[4]

Because the parties were so far apart on most issues, the British decided that traditional negotiating patterns of establishing opening positions and trading concessions toward an acceptable midpoint would not work. Instead, Carrington would take the initiative, offer at each stage an outline plan, amend it as much as he believed practicable on the basis of the responses of the other delegations, and then present a final detailed proposal to which he would, if necessary, demand a response. Lord Soames, who was to play a major role in Rhodesia after the conference, has written:

> The role adopted by the British Government was always to guide the negotiations from the general to the particular. At every stage we sought to introduce—just as we had done before, during and after the Lusaka conference—statements of broad principle to which it was very difficult for the delegations themselves, or their supporters outside Lancaster House, to take exception. We then proceeded step by step to deduce the logical

consequences which we argued had to flow from assent to these statements. Here again we were learning from the experience of the past, which strongly suggested that the introduction of too much detail too soon, however worthy the intention with which it was done, in practice led debate into side issues and enabled those who were so inclined to evade the major questions which *had* to be settled before a solution could be in sight.[5]

Underlying the tactics was the skepticism shared by Carrington, Mrs. Thatcher, and most of the FCO professionals about the conference's prospects. A negotiated settlement acceptable to both the Patriotic Front and the Muzorewa government was viewed as unlikely—surely to be sought after, but probably impossible to attain. This pessimism promoted a more "realistic" prognosis of a settlement acceptable to the bishop—perhaps including Nkomo, who might be induced to split from Mugabe—which could be implemented and gain support from important sectors of the international community. The "second-class solution," as it was called, became much more than Britain's acknowledged fallback position. In practice, Carrington's willingness to accept such an outcome became his principal negotiating lever to be used on the other delegations: for one a hope, for the other a threat.

The Salisbury Team

The delegation led by Bishop Muzorewa was a heterogeneous group, united in some objectives, but severely divided elsewhere. The official twelve-man team represented Salisbury's politically hybrid government of national unity. It included black adversaries of the bishop such as the Reverend Ndabaningi Sithole, the first president and still claimant to the leadership of Mugabe's Zimbabwe African National Union (ZANU), as well as four white cabinet members, led by Ian Smith.

The forty-four-year-old bishop was himself a contradictory character, embodying both a sincere dedication to majority rule and an egoistical desire for personal enhancement. His story is not an atypical one of the colonial Africa of his time: the son of peasants, the hardworking "mission boy" is trained to be a Methodist pastor, is later sent to small church colleges in the U.S. South, returns home to further religious endeavors, and, in rather short order, becomes the first Rhodesian African to be consecrated a bishop of the United Methodist Church. In 1971 Muzorewa's churchly activism caused him difficulties with the Smith regime, which banned him from entering African tribal reserves. The notoriety gained in the confrontation propelled him into the leadership of the African National Council (ANC), formed by the contending nationalist movements to oppose the Smith-Home constitutional proposals.

After the Pearce Commission submitted its negative report, Muzorewa determined not to return to strictly clerical pursuits and from 1972 onward presented himself as the most legitimate claimant to the role of leader of a unified African nationalist movement. For a brief period in 1974 and 1975 the other nationalist leaders again agreed upon Muzorewa as the head of the ANC in failed negotiations with Smith. A diminutive man, no more than five feet in height, Muzorewa commented at the time of his selection as the compromise figurehead leader that "I feel like the tallest man in the world."[6] But ANC unity was short-lived and the organization divided into its separate entities after the Victoria Falls conference between Smith and the nationalist leaders produced no movement toward majority rule. Muzorewa led his own delegation to the Geneva Conference of 1976. By late 1977 he was willing to enter into negotiations with Smith, which led to the internal settlement of March 1978.

A vain man with a passion for flashy clothes, including blue patent leather boots and multicolored tribal robes of his own design, Muzorewa was not a skilled politician. In the internal settlement negotiations he and the other domestic African leaders at the table consistently gave away bargaining chips to the white regime. His few months of rule as prime minister after the April 1979 elections had been lackluster, replete with indecisive half steps that did not please the African majority, but, at least from his point of view, did not alienate the white military and administrative structure that kept him in power.

Britain's long history of dealing with Smith, whom they viewed as a devious negotiator, did not fully prepare them for the bishop. There had been a certain method in Smith's craftiness, but Muzorewa was neither as secure nor as skilled as his white predecessor. His ego and his weakness as a negotiator opened opportunities that the British were to exploit during the conference, but made the settlement effort that much more difficult. Confidantes of Carrington reported after Lancaster House that although the Foreign Secretary truly disliked Smith, no other delegate so enraged or frustrated him as the vacillating bishop.

With only a few exceptions, the African members of the Salisbury delegation were not as experienced or as skilled as many on the Patriotic Front team. White civil servants provided what little administrative backbone the delegation possessed. Observers noted from the outset that the delegation, which did not meet formally as a group until arriving in London, lacked cohesion and bureaucratic competence.

Muzorewa came to the conference buoyed by the results of the April 1979 election, which seemingly demonstrated that the mass of the African majority favored him as their leader. However, he was also aware that the large vote had been a popular expression of the grass roots' desire for peace. As the war had grown in intensity, it had been the African population, particularly rural dwelleres, who had suffered the most. Between 20,000 and 50,000 (estimates vary widely) had been killed by the guerrillas or the Rhodesian security forces. Another million, or one out of seven of the country's black population, had been displaced. They lived as refugees in the surrounding countries, as urban squatters, or as usually unwilling residents of the government-run "protected hamlets." Bringing peace to the country had been Muzorewa's overriding campaign theme, but in his few months in office the war had not abated. A much heralded amnesty program for guerrillas had shown itself to be a dud. The bishop was being told by the international press, if not by some of the sycophants who surrounded him, that he was losing popularity.

For the bishop's delegation the central objective of the conference was obtaining recognition from Britain of its legitimacy as a government and, flowing from that, the lifting of international sanctions. The same economic malaise that had propelled Smith into the internal settlement now engined the Muzorewa regime's search for respectability. An end to sanctions would mean greater funds available to the government for prosecution of the war and, of greater concern to Muzorewa than it had been to Smith, for social programs of development among the African majority. Moreover, a duly recognized government would be able to call on friendly nations for military assistance in combating the guerrillas. The ability to acquire additional sources of foreign support, apart from that provided by the South African government, was a major goal of the Muzorewa regime and could be attained only after the lifting of sanctions and formal granting of independence. Those goals could, in turn, be met only by acting in cooperation with Britain to find a suitable settlement.

Of perhaps greater importance than the deteriorating economic situation was the growing awareness among Rhodesia's white intelligence and military professionals, some of whom were members of the delegation's official staff, that the war had reached a stalemate. The guerrillas could not win. But the regime's forces were being stretched thinner as fewer "Europeans" were inclined to be the "last white to die for a black government." Smith himself undoubtedly recognized the economic and military realities, but was unwilling to come to terms with them. This prompted a duality during the conference in which he cast negative votes on crucial issues to maintain his record as a defender of white rights, but declined, despite press speculation to the contrary, to rally his supporters at home against the conference's proceedings. And protest he could have, because most of his traditional backers were not aware of the country's dire straits. True, white farmers lived in armed camps, if they had not already been driven off their land, but for the majority of urban whites the inconveniences of the war—gas rationing, reserve call-ups, and curfews—had come gradually and been adjusted to. Radio Rhodesia pumped out the party line, trumpeting the death of terrorists, and the vaguely pre–World War II British air of dog shows, afternoon teas, and rugby prevailed. The white community as a whole maintained an almost surrealistically high morale.[7]

Though united in their goals of recognition, the lifting of sanctions, and the end to, or at least scaling down of, the war, the Muzorewa delegation was nonetheless divided. Smith and several other whites had no respect for the bishop, whom they viewed as indecisive and no match for the wily British. The blacks, on the other hand, harbored long-standing distrust of the whites and had a private agenda of their own. They wanted to use the conference as a means of eliminating many of the entrenched and inherently racist provisions of the constitution they had been obliged to accept during the internal negotiations of 1978. It would be the constitutional issue that the British would tackle first, and in that debate the delegation's white members were to abandon Ian Smith, providing for the conference's initial drama.

However, once that hurdle had been overcome, Carrington would still have to convince Muzorewa to relinquish his hold on power. The Lusaka accord demanded free elections "properly supervised under British Government authority." The British recognized that they would have to assume actual control of the government of Rhodesia in order to be able to run the elections. Lack of U.K. control would subject the electoral process to the same charges of farce and fix that had been leveled at the April 1979 voting. Muzorewa may not have realized when he arrived in London that he would have to stand down, but Carrington knew what had to be done. Gaining the bishop's acquiescence in his own abdication was, British officials later said, their single most difficult task of the conference.

The tension within Muzorewa's delegation was not to emerge visibly until a few days after the conference's start. At the outset it was papered over with an optimism born of a simple and seemingly obvious fact: Tory support for Muzorewa, when coupled with the likely obstinacy of the Patriotic Front, would bring the conference to a rapid dissolution. The British would have no option to—and indeed many thought would welcome—coming to a separate agreement with the bishop. Carrington's second-class solution was Salisbury's first choice.

The Patriotic Front

The delegation led by Nkomo and Mugabe also assumed a British bias in the bishop's favor and also doubted that a comprehensive settlement involving them could be achieved. Given this suspicion and skepticism, why did the Patriotic Front accept the British invitation? In some measure they were like the addicted gamblers who know that the local poker table is fixed, but continue to play because it is the only game in town. However slim the chances of a fair deal might be, the conference nevertheless presented an opportunity that could not be passed up. There were other reasons as well.

Though Nkomo and Mugabe, more so the latter, believed that a military victory was still possible, they recognized that the prospects for an early triumph in the field were small. As in the Muzorewa delegation, it was those closest to the fighting, particularly the head of ZANU's army, General Josiah Tongogara, who had the clearest understanding of the disastrous nature of the war and who would argue during the conference for a course of moderation and compromise. Also, the Patriotic Front's inability to disrupt the April 1979 elections demonstrated that the Rhodesian regime, still orchestrated and controlled by whites, was undeniably functional and resilient.[8]

Another important factor in the Patriotic Front's decision to attend was the influence of the Front Line states, notably Zambia, Tanzania, and Mozambique. The prestige of the first two was at stake as they had played key roles at Lusaka.[9] More importantly, their economies, particularly those of Zambia and Mozambique, were suffering a horrendous toll as a result of the Rhodesian war.

According to the Commonwealth Secretariat, the principled decision of Zambia's President Kaunda to observe sanctions against Rhodesia had cost his country over US $750 million by 1977.[10] Even allowing for a certain amount of sympathetic expansion that might include in that figure the results of a dramatic decline in world copper prices or the economic losses produced by the country's ill-managed state enterprises, the overall effect was disastrous.

In Zambia public disaffection with the war was palpable and the resentment was increasingly directed not only toward the Smith regime and Nkomo's troops quartered there, but to Kaunda's government as well. Shiva Naipaul, a gifted observer, visited Lusaka in early 1978 and wrote:

> The Rhodesian question provokes irritation, not altruism. The irritation was rampant in the young clerk who complained to me about the steep rise that had occurred in the cost of living. . . . He did not attribute his poverty to the fall in the price of copper—which earns for Zambia the bulk of its foreign exchange—but laid the blame on the support the government was giving to the "liberation struggle." "The ministers are all right—they are rich men. The freedom fighters are all right—they get free food. But what about people like us who are not ministers and not freedom fighters?"[11]

President Kaunda would send his political "fixit" man, Mark Chona, to be his personal representative in Lancaster House's environs. Kaunda himself would visit London once during the conference and play a helpful role in bringing about a compromise at a critical juncture.

Shortly after assuming power in 1975, the Mozambican government of President Samora Machel closed its borders to the Smith regime, robbing Rhodesia of its access to the sea through the ports of Lourenço Marques (renamed Maputo) and Beira. Sanctions were estimated to be costing Mozambique US $100 million a year. However, Mozambique, as a former Portuguese colony, had not been represented at the Lusaka Commonwealth Conference, and, at the beginning of the Lancaster House Conference, it was unclear to many British and U.S. observers what role the Mozambicans would play.[12]

In fact, the Mozambican observers, led by Fernando Honwana and Jose Luis Cabaco, were to be the most instrumental of the Front Liners in pressing the Patriotic Front toward a settlement. Their government had been told by their own spies and soldiers fighting alongside ZANU inside Rhodesia that in any fair electoral test the Patriotic Front would be an overwhelming victor.[13] It was an assessment Mugabe and Nkomo shared as well for, like Bishop Muzorewa, they were convinced of their own popularity among the African masses of Zimbabwe.

Finally, and perhaps the most significant motive force compelling the Patriotic Front to attend the conference, was its desire not to give the British a clear path to recognition of the Muzorewa regime. Just as the bishop's delegation was intent on sticking it out until the Patriotic Front stormily departed, Nkomo and Mugabe were committed to not giving Muzorewa and Carrington (whom they viewed as a team) such an opportunity. As the *Economist* noted at the conference's outset in an otherwise gloomy piece of prognostication, both Zimbabwean delegations "face traps the avoidance of which could just condemn this conference to 'succeed.' "[14]

Despite a general unity of perception, the Patriotic Front, like the Muzorewa delegation, was an uneasy coalition. Almost two decades of hostility and rivalry separated Mugabe and Nkomo. The former had been one of the leading young Turks who challenged the latter's leadership of the Zimbabwe African People's Union (ZAPU) in the early 1960s—a challenge that ultimately led to the ZANU-ZAPU split. Prior to the Geneva Conference of 1976 the Front Line states had compelled the two factions to unite as the Patriotic Front for negotiating purposes, but the alliance was a fitful one. The two armies did not coordinate activities and, on occasion, fought each other. Mugabe's suspicion of Nkomo was fueled by the ZAPU leader's continuing flirtations and open lines of communication with Smith, with the governments of Britain, the Soviet Union, the United States, and South Africa, and with certain multinational corporations and various others—all, in Mugabe's view, unpalatable associates.

Mugabe only had to read the London press to confirm his suspicion that the British and the Rhodesian whites still maintained the hope that Nkomo might be induced to divorce himself from Mugabe and go along with a second-class solution. The British would discreetly discuss this with Nkomo during the conference and again during the transitional period. However, as events proved, the potential rewards to Nkomo for divorcing himself from Mugabe were never great enough to outweigh the problems inherent in such a move. In order to gain the political supremacy he coveted in an independent Zimbabwe, Nkomo, a member of the minority Matabele tribe, recognized that he would have to maintain a political alliance with Mugabe's ZANU, which drew its strength from the majority Shona ethnic group (80 percent of the population). But there were times when he was tempted, and the possibility of a split was a constant factor in intra–Patriotic Front negotiations.

The sixty-two-year-old Nkomo and the fifty-five-year-old Mugabe were similar in some ways. Like other Zimbabwean leaders of their generation they owed their early education to white missionaries. Both had spent more than a decade imprisoned or otherwise confined by the Smith regime. And, of course, both leaders were firmly committed to African majority rule.

But there were also significant differences in the style, outlook, and relative organizational strengths of the two men that would influence their actions at Lancaster House. Nkomo had first entered public affairs as a labor leader and he still maintained a sort of back-slapping, hail-fellow-well-met persona. His Falstaffian physique conveyed an air of avuncular pleasantness, which was somehow comforting to the wide variety of individuals and governments that found him easier to approach than his leaner, more austere colleague. Mugabe had been a schoolteacher. During his long years in prison he had devoted much of his time to study and reflection, obtaining in the process three university degrees by correspondence. He was more introspective, less outgoing, more intellectual than the ZAPU leader—cool where Nkomo was warm.

The two men also had differing approaches to the nature and goals of the war they were directing. Heavily influenced by Marxist-Leninist and other revolutionary thinkers, Mugabe viewed violence as an inescapable and integral element of revolutionary struggle, necessary not only to defeat the enemy but also to politicize the fighters and the masses and to prepare them for the kind of major societal restructuring a new state would bring about. For Nkomo, violence was a tactic, but not necessarily an ennobling enterprise. Unlike Mugabe's, Nkomo's ideological pronouncements about the economy of an independent Zimbabwe were fuzzy. His occasional espousals of state socialism were not given serious credit by most observers, who were well aware of Nkomo's own capitalist endeavors.[15]

The relation of each man to his political and military organizations also differed. Nkomo was ZAPU. He was its first and only leader, and its army had been created by him after his position of authority had been well established. Perhaps reflecting the kingly tradition of the Matabele, who invested their chiefs with more power than did the Shona, Nkomo tended to rule his delegation by fiat, tolerating little debate and permitting less dissension. Mugabe, on the other hand, owed his position of power to ZANU's military forces. The organization's history was replete with internecine conflict, some of it quite violent. It was not until early 1976 that ZANU's military leaders threw their weight behind Mugabe as the organization's principal political force. The military continued to play a considerable role in ZANU's deliberations as evidenced by the influence that was to be wielded at Lancaster House by the army commander, General Josiah Tongogara.[16]

Given ZANU's fractious nature, Mugabe had become accustomed to ruling by consensus only after lengthy debate. This was necessary to keep a balance between fire-breathing radicals such as the party's Secretary General, Edgar Tekere, and more pragmatic leaders, including legal advisor Simbi Mubako, Tongogara, and intelligence chief Emmerson Munangagwa. The convoluted relations and tensions among the three major Shona subtribal groupings may have also played a part in determining Mugabe's negotiating style. As a Zezuru, he had to tread carefully around the sensitivities of the more numerous Karangas, whose members formed the bulk of ZANU's fighting force and its military leadership.

What had united Mugabe and Nkomo in past negotiations was a consistent commitment to the proposition that their forces, both political and military, had earned the right to exercise power prior to the selection of an independent Zimbabwe's first government. This was less a question of pride than of practicality. Only by controlling the mechanisms of government, including the police and military forces, could the Patriotic Front be guaranteed that they would have a fair opportunity to elicit the popular support they knew existed in an electoral contest. This insistence on control, total or at least predominant, in any transitional period was the mirror image of the bishop's intention to maintain his power undiminished. Convincing the Patriotic Front to relinquish their insistence on transitional power would prove a wearying experience for Carrington and compel the British into a greater transitional period involvement than they had initially contemplated.

Thus, of the three delegations that assembled on September 10 at Lancaster House, only the British were able to present a unified stance. The other two teams were burdened with internal dissension and distrust. About the only things that the Patriotic Front and the Salisbury delegations had in common, it seemed, were their mutually exclusive claims to power, their convictions that they each enjoyed the support of the African masses, and the firm desire not to be the first to leave the conference. On this narrow base Lord Carrington began to construct his Rhodesian settlement.

4
Overture and the Constitution

The Lancaster House Conference was a three-act play, or, better put, a one-act play performed three times, but with enough variety and tension so as not to rob each performance of its drama. Each dealt with a distinct area of the settlement— the constitution, the transitional arrangements, and the cease-fire—and lasted about one month. Most of the performance was enacted outside of the conference hall. At each stage the British were able to obtain early Muzorewa approval of their proposals and couple that with Carrington's perceived willingness to pursue a second-class solution, to obtain the Patriotic Front's acquiescence. In retrospect, the three acts take on an almost formalized predictability, but at the time few could be sanguine of success. An overture, containing many themes to be subsequently amplified, preceded the first act.

Overture

Appended to Carrington's invitations of August 14, 1979, to Muzorewa and the Patriotic Front leaders was an outline of constitutional proposals on the basis of which Britain would be prepared to grant independence.[1] The eleven proposals called for universal adult suffrage, a parliamentary form of government, a constitution with a justiciable bill of rights, an unspecified number of parliamentary seats reserved for whites, and a civil service and military establishment responsible to elected officials. Parliament would also have the right to amend the constitution under procedures similar to those contained in other independence constitutions granted by Britain.[2] The promise of political control over public officials and a traditional form of constitutional amendment was clearly designed to demonstrate Britain's intention to use the conference to eliminate the most objectionable elements of the internal settlement. But the proposals' broad phrasing and the invitations' acknowledgment of British willingness to consider other formulations, if acceptable to the parties, rendered them noncontentious. Of perhaps equal interest to the parties was Carrington's decision to chair the conference himself, signifying that Britain was not hedging its bets or political commitment.

The invitations met with predictable responses. Though still fuming about Mrs. Thatcher's "sell-out" at Lusaka a week earlier, Muzorewa signaled within a day of receipt of his invitation his intention to attend. The Patriotic Front took a bit longer to respond. In the interim, Nkomo and Mugabe publicly repeated their preconditions for talks, which included the liquidation of the Muzorewa-Smith regime and the disbanding of the Rhodesian army—demands that were not taken seriously by Carrington. After meetings with Angolan President Neto and Tanzanian President Nyerere, the Patriotic Front's leaders accepted the invitation on August 20. The conference was scheduled to begin on Monday, September 10.

However, even after responding, Nkomo and Mugabe tried to maximize their negotiating position by seeking assistance to undermine the Lusaka accord. At the early September meeting of the Nonaligned Summit in Havana, leaders of the Front Line states became incensed when they discovered that the Patriotic Front was seeking support for a resolution that would condemn the agreement reached in Lusaka. Nkomo and Mugabe specifically objected to the Lusaka conference's implicit rejection of the guerrillas' demands for a transfer of power to them prior to elections. Presidents Nyerere and Machel were particularly aggravated and fearful that the Patriotic Front's actions would sabotage the upcoming London meeting.[3] They reportedly told Nkomo and Mugabe in Havana that they must attend the London conference on the basis of the Lusaka agreement. Failure to do so would constitute cause for withdrawing Front Line support from the war effort. On the other hand, if it became apparent at Lancaster House that the British were simply staging a new conference to mask an already reached agreement with Muzorewa, Front Line support for the fighting would be increased. Faced with that ultimatum, which would not be the last time a similar threat would be issued, Nkomo and Mugabe acquiesced.

Muzorewa's September 7 departure from Salisbury coincided with a Rhodesian Security Force announcement that it had just completed a three-day raid into Mozambique in which it claimed 300 guerrillas had been killed and nine bases and six road and rail bridges destroyed. Mozambique army bases near guerrilla targets also came under attack. The operation was designed to demonstrate the regime's continued military might and to impress upon the Front Line states the heightened costs of the war.

As the delegates arrived in London their own pessimism was reinforced by the prevailing mood of the London press, which, briefed by the FCO, focused on the distance separating the parties. The *Financial Times* noted that "there will have to be compromise, but if history is any guide, there is little hope of that."[4] Other articles reported British expectations that Nkomo and Mugabe, or Mugabe alone, would walk out of the conference and the probability of the Thatcher government's pushing for a second-class solution.

In Lancaster House's ornate meeting hall, the British had taken some care in the seating and labeling arrangements, consciously referring to "Mr. Mugabe's and Mr. Nkomo's delegation" and "Bishop Muzorewa's delegation" in order to avoid the wrangling that would have been prompted by more formal appellations, such as "the government of Zimbabwe-Rhodesia." Nevertheless, the Patriotic Front argued that the Salisbury delegation should be seated as part of the U.K. team: as Rhodesia remained a British colony, its temporary governors were simply agents of the Crown. Carrington welcomed the Patriotic Front's opening gambit as an opportunity to flex his own muscles and to demonstrate the no-nonsense school of conference management which he and his staff previously had decided upon. He brusquely refused to entertain the complaint and let it be known that such procedural haggling would be taboo. Having probed for a soft spot, Nkomo and Mugabe found none and pulled back.

Carrington's upbeat opening speech on September 10 placed the gathering within the context of Britain's decolonizing past. "As the constitutional authority for Southern Rhodesia," he said, "the United Kingdom intends to take direct responsibility for the independence constitution." He repeated his willingness to entertain alternative formulations if agreed to by the parties.[5] He then adjourned the plenary, asking the other two delegations to present their views on the following day. That evening Nkomo and Mugabe boycotted a reception hosted by the British, telling the press that meeting socially with the Salisbury delegation would be like Carrington's going to a party with the killers of the recently slain Lord Mountbatten.

At the next morning's plenary Nkomo read a prepared statement asserting the exclusive role of the Patriotic Front in forcing the Rhodesian regime to the negotiating table where "the two decolonizing forces . . . the Patriotic Front representing the people of Zimbabwe . . . and Britain . . . here asserting her diminished legal authority" would complete the decolonizing process. He labeled Carrington's constitutional proposals as "too vague" and rejected the British constitution-first agenda. Instead, he argued that as the transitional period would be key to the holding of free and fair elections, it must be discussed first.[6] On the following day the Patriotic Front formally presented its amended agenda, urging sequential discussion of the transitional administrative arrangements, including status and treatment of forces; the constitution for the transitional period; ceasefire arrangements; and, finally, the independence constitution.

In his opening remarks Muzorewa placed his government's current constitution on the table as its negotiating document, arguing that it satisfied the six principles and "that nothing should now stand in the way of our Government of Zimbabwe-Rhodesia being granted their rightful recognition." However, he grudgingly and indirectly acknowledged that some change might be possible:

> We require to know clearly and categorically what more your government requires from us before you will remove sanctions and grant recognition to our government. Thereafter, a firm commitment in specific terms from your government that it is prepared to support our government to the fullest extent, that sanctions will be lifted, and that recognition will be granted.[7]

Implicit in Muzorewa's presentation was his delegation's view that they had come to London solely to negotiate with the British. The Patriotic Front's presence was irrelevant.

The challenge by Nkomo and Mugabe to the British agenda was considerably more serious than the flareup about seating arrangements or the bishop's initial attempt to force Carrington to set the outer limits of what would be required of Salisbury. The demand purposefully struck at the heart of the U.K.'s plan to move step by step, beginning with the constitution. Acceptance of the alternate agenda was unthinkable to Carrington and his response was therefore predictable. However, how he and his staff handled the challenge adumbrated a style they were to adopt at various points in the conference—a hard public posture accompanied by flexibility in camera. At the September 12 plenary Carrington presented a formulation—previously cleared by his staff with the other delegations—stating that although the British agenda was inviolable, all the topics on the Patriotic Front's list would be discussed, but only after the constitution had been agreed upon. With that pronouncement both the Patriotic Front and the British could claim a minor victory. However, in reality, Carrington's was far greater because he had preserved the essence of his negotiating tactic.

The Constitution

The first full act of the Lancaster House Conference was to differ from the following two in that it featured an open breach within the Muzorewa delegation and witnessed hard, relatively public bargaining by the British with recalcitrant members of the Salisbury group. Otherwise, the constitutional negotiations resembled what followed in Carrington's basic tactical approach, the involvement of supporting actors, and the impact of important outside events—in this case the Conservative Party Congress—which lent an air of tension to the conference proceedings.

As soon as the agenda dispute was settled, Carrington distributed a thirteen-page summary of constitutional proposals that amplified the principles accompanying the invitations and made clearer the British intention to modify significantly the more offensive elements of the internal settlement constitution. The proposals seriously alarmed some of the white members of the Salisbury delegation. With the presentation on September 14 of the Patriotic Front's draft constitution, three documents were thus on the table. Not surprisingly, and consistent with preconference planning, Carrington insisted that the British draft form the basis of discussions.

Mindful of white concerns, the U.K. document contained numerous provisions designed to protect that minority's interests, including reserved parliamentary seats, an extensive bill of rights enshrining traditional Western freedoms, and a proposal that certain sections of the constitution dealing with such crucial topics as the judiciary, the legislature, and the amendment process itself be entrenched for a period of years. Pension rights for public officials would be honored, and a special incentive scheme designed at the outset of the internal settlement to keep white civil servants working for a black government would be continued. The British draft differed markedly from that presented by the Patriotic Front, which was singlemindedly egalitarian and simply offered all citizens "security and not privilege . . . and equal rights without discrimination." The latter plan made no provision for special parliamentary seats for whites, for entrenched constitutional elements, or for the protection of private property or pension rights. It also contained restrictive citizenship provisions. The

U.K. draft would grant automatic Zimbabwean citizenship to any Rhodesian and indefinitely permit dual nationality.[8]

The battle now joined, the British concentrated the bulk of their forces on the Muzorewa team, which was already badly split over the new U.K. draft. Carrington was aware that there was little in his proposals that could offend the delegation's African members. In fact, his officials had already been told by two of the bishop's confidantes, Deputy Prime Minister Silas Mundawarara and Muzorewa's private secretary, James Kamusikiri, that Muzorewa welcomed the proposals that limited white power.[9] It was not a position he could stake out publicly. However, in private the bishop's men tried to convince their white colleagues, prompting one Salisbury African delegate to tell the press that "there are two sets of London negotiations. One here at our hotel and the other at Lancaster House."[10]

It fell to the British to make the most persuasive arguments in favor of the Salisbury delegation's acceptance of Carrington's draft. The opposition was led by Smith, assisted by Chris Andersen, Rhodesia's most successful and abrasive trial lawyer, who served as Muzorewa's Minister of Justice.[11] Smith and his allies were buoyed by the advice they received from Tory right-wingers to hold firm. Their Tory sources said Carrington would be unwilling to risk a revolt at the Tory Congress scheduled to begin October 10, and that he certainly would not be able to collect sufficient Conservative votes to guarantee passage of new sanctions when the yearly vote on that issue would arise on November 15. The right-wing arguments were bolstered by the arrival in London for a brief visit of two members of the staff of conservative U.S. Senator Jesse Helms, who told Smith that the U.S. Congress would defy President Carter and lift sanctions in the near future.[12]

The British mustered several telling arguments against the "hold firm and reject" position. FCO officials explained to Smith and the other whites that, if agreement had not yet been reached by November 15, Mrs. Thatcher would not have to risk a parliamentary battle over sanctions, as fully 90 percent of the sanctions were authorized by other legislation and would remain in force until specifically repealed.[13] Having weakened the sanctions issue as a negotiating weapon, Carrington's team then stressed that as no element had so galvanized world opinion against the internal settlement as the constitutional provisions protecting white privilege, these provisions had to be altered. Moreover, those issues of particular importance to whites—property, political representation, and pension rights— were covered by the British draft.

The British also took a standard white Rhodesian line—a constitution is only a piece of paper that can be torn up at any time, as had happened in several other African states— and turned it on its head. True, they said, a constitution could easily be destroyed. Therefore, why haggle over the number of white seats or blocking mechanisms? What would really matter in the long term would be the spirit of cooperation between the races. The place to begin creating such a spirit was Lancaster House. Finally, the most telling British debating point was an insistence that without Salisbury's acquiescence to Carrington's proposals, no settlement, not even a second-class solution, would be feasible.

The British arguments were reinforced by David Young, Rhodesia's Permanent Secretary of Finance and a member of the Salisbury support team of white officials. Young had masterminded Rhodesia's growth during the early sanction years and later expertly balanced the competing needs of fighting a war and keeping the rest of the government functioning. His facts and figures on the state of the economy and the need for a quick settlement carried weight with the three white delegates—David Smith, a respected businessman who had always remained somewhat aloof from the hurly burly of Rhodesian Front politics; Rowan Cronje, an Afrikaner; and Andersen—who, with Ian Smith, were the Salisbury delegation's voting white members.

Taken together the arguments were persuasive. On September 21, the Muzorewa delegation voted eleven to one to accept the British draft. Smith cast the lone negative vote. Publicly, Carrington called the Salisbury decision a major advance. Privately, FCO officials told the press that they saw less than a 20 percent chance of a full agreement. The pessimism may have been genuine, but, as in so much of the British manipulation of the press, it served a dual tactical purpose—reinforcing with the Muzorewa delegation that it had acted wisely and at the same time warning the Patriotic Front that further delay or obstructionism would be counterproductive.

While meeting with the Muzorewa delegation, British officials had also undertaken parallel private negotiations with the Patriotic Front's leaders. At a plenary session on September 24 Nkomo and Mugabe accepted the proposition of 20 percent reserved representation for whites in the lower house. This was a major concession on Mugabe's part. He realized that a guaranteed twenty seats in a 100-member parliamentary body could place the whites in a kingmaking role if the African vote were seriously fractionated. (The language of the constitution when finally promulgated by the British prohibited the white members from joining with a single African party to form a majority and thus constitute a government. In theory, however, that language would not have prevented the twenty whites from combining with two or more African minority parties to form a government.) The prospect of the reserved seats was not as troubling for Nkomo, who wished to close no doors to potential coalition partners. Although at this time he wanted ZANU and ZAPU to contest the election as one party, Nkomo may have realized that Mugabe would not be inclined to do so. As events developed, the two parties ran separately, and Mugabe's overwhelming victory precluded the issue of white coalition with minority parties, including Nkomo's, from arising.

Front Line representatives exerted considerable pressure on the Patriotic Front to accept the 20 percent solution.[14] President Nyerere had already expressed his approval of the draft when it was shown to him by the British before they had tabled it at Lancaster House. The Front Line representatives argued that some demonstration of Patriotic Front flexibility was called for, that the constitution could be amended at a later date, and that should the conference break down over this issue, the British would have no alternative but to recognize the Muzorewa regime. Nkomo and Mugabe finally acquiesced, but were not willing to make any more concessions. There were to be no further plenary sessions for almost two weeks as the British continued their parallel bilateral negotiations, which yielded little visible progress. Carrington finally resumed plenaries on October 2 and on the following day presented his third and final draft constitutional proposals, giving both sides until Monday, October 8, to convey to him a firm yes or no.

In presenting the new draft, the British referred to the thirty-four-page document as a fair product of arbitration, adding what was reasonable from the views of both delegations to its own thirteen-page proposal of September 13. In fact, a careful reading of the new document revealed that while the Patriotic Front had been debating grand policy with the British, Salisbury's white contingent, inspired by Smith and field marshalled by Andersen, had successfully pressed for further advantage.[15] The amplified bill of rights now proposed that compensation paid by the state for the deprivation of property would, under normal circumstances, be freely remittable out of Zimbabwe. Further, "property" was defined to include pension benefits, in effect converting pensions from a payment for services rendered to an inalienable right. Other provisions also spoke to white concerns by strengthening the autonomy of the attorney general and by prohibiting the government from preventing any person or group from establishing a school or from sending a child to the school of his choice. The new draft also entrenched the reserved seats for a period of at least seven years, amendable after that time only by a 70 percent vote of the House of Assembly. Provisions of the bill of rights would be entrenched for ten years, amendable during that time only by a unanimous vote of the lower house. There were a few provisions that conceivably could have been viewed by the Patriotic Front as bows in their direction, but, in sum, the document revealed that their haggling had produced little in the way of British concessions to them.

Satisfied that they had obtained the best constitutional deal likely, on October 5 the Salisbury delegation voted once again eleven to one to accept the expanded British draft. Andersen, though voting in favor, cast Smith's negative proxy vote as Smith prepared for a brief visit home. In announcing the vote, Muzorewa also added that he would accept new elections: "Although we feel another election is unnecessary and unfair to Zimbabwe-Rhodesia, we will do so in full confidence that our people will reaffirm their desire for and commitment to genuine democracy."[16] The bishop's statement was a major step for a man who had only recently contested one election and had come to London insisting that Lancaster House would simply be a constitutional conference that would not entail the need for new voting. The timing of the Muzorewa delegation's vote was critical. In part, it may have been designed to discourage Smith from rallying opposition to the London proceedings while he visited Rhodesia. More importantly, the bishop had now placed himself in the position of accommodating Carrington and accepting the key provisions of the Lusaka accord. Surely, given continued Patriotic Front delaying tactics, Carrington would have no option but to accept the pressure that would certainly be generated at the following week's Tory Congress for a second-class solution.

The Salisbury decision placed increased pressure on Nkomo and Mugabe. Over the weekend of October 6–7 the leaders met to plot tactics and prepare a response to Carrington's deadline. Front Line and Commonwealth representatives reportedly urged caution and, indeed, the Patriotic Front's press spokesmen rejected the possibility of conference breakdown and called for further negotiation. At the October 8 plenary a "yes" or "no" response, was withheld, but a tentative "maybe" was offered. The Patriotic Front presented a detailed criticism of Carrington's latest constitutional proposals, objecting to numerous provisions including citizenship and pension rights, unamendability of the declaration of rights, limitations on the ability to recover land from which Africans had been dispossessed, the remittability of land compensation, and several other elements that they viewed as unnecessarily restrictive of the freedom of the government of an independent Zimbabwe. Having listed their criticisms, the delegates concluded their document with "we have detailed our major reservations. We propose we now proceed to discuss the second item on the agenda, the interim arrangements."[17]

Once again, the Patriotic Front was posing a serious challenge to Carrington's conference tactics. To accept the idea to "agree to disagree and then proceed" would have clearly weakened the step-by-step approach, angered the Muzorewa delegation that had, after all, played by Carrington's rules, and made further negotiation more difficult as each of the other parties would withhold final approval to a given step while it sought additional bargaining advantage. The Patriotic Front's gambit was a tactically logical move: if transitional and other arrangements acceptable to them could be agreed upon, and they probably still thought this to be unlikely, the pressure on the British and the Muzorewa delegations to accept their constitutional views as the final piece of the jigsaw puzzle would be measurably increased. Carrington had to hold firm. In a brief plenary on the following day he rejected the Patriotic Front's suggestion and demanded a formal decision from Nkomo and Mugabe by October 11. Presenting a hard line was an imperative for Carrington as he left London to attend the Conservative Party Congress in Blackpool.

In retrospect, the congress's easy treatment of Carrington should have been predictable: internecine warfare so soon after having assumed power would have been senseless for the Tories. In addition, the Tory leadership could persuasively argue against precipitant action that would prejudge Lancaster House. The party machinery had purposely selected for debate the mildest of the forty-two Rhodesian resolutions submitted by local chapters, one that simply called for independence and the lifting of sanctions as soon as possible. Carrington was thus able to support the resolution, asserting that one way or another, Lancaster House held promise for a rapid solution of the problem. In a well-reasoned speech, he also informed the congress that Britain's job would not be finished with Lancaster House: elections would have to be supervised, British officials would have to be on the scene. His suggestion of an activist role for Britain struck the right chord of resolve for a congress that was to support Mrs. Thatcher's call for "no

U-turn" in her stringent economic policies. With Mrs. Thatcher demonstrating her support by appearing on the podium with Carrington while he spoke, he received a standing ovation. Julian Amery was permitted to address the congress and criticized Carrington's approach, but he neither sought nor could have succeeded in promoting a right-wing ruckus. The *Guardian* expressed the view of most observers when it commented that Carrington had been given a "blank cheque."[18]

On his return to London Carrington proceeded to cash the cheque. In his absence the Patriotic Front had presented a tough public posture with Mugabe echoing Mao:

> If this London conference reaches no decisions, we will despatch our military men back to Africa. This means the intensification of the struggle. We can win without Lancaster House. That is a certainty. Of course, we would welcome a settlement. But we can achieve peace and justice for our people through the barrel of a gun.[19]

After a bilateral meeting in which Nkomo and Mugabe refused to budge, Carrington adopted a tough public posture of his own, indefinitely adjourning the conference. At the same meeting, following his practice of a more flexible private approach, he offered an olive branch to the Patriotic Front, whose criticism of the British plan had increasingly focused on the issue of compensating whites for land that the Patriotic Front argued had been stolen from its original African owners. Carrington told Mugabe and Nkomo that Britain would be prepared to grant financial assistance for land resettlement and redistribution schemes that an independent Zimbabwean government might undertake. He did not commit the U.K. to a specific sum, but his intent was clear: the constitution, to be acceptable to the Salisbury delegation, must provide for compensation to whites for land taken by the new government, but Britain would be prepared to shoulder some of the financial burden. The British offer was pocketed by the Patriotic Front's leaders, and did not lead to an immediate change in their stance.

Carrington waited a few days more while his officials tried to find common ground with the Patriotic Front, but on October 15 he publicly increased pressure on Nkomo and Mugabe while privately sweetening the land compensation offer. Carrington scheduled a press conference for that day, a rare move in that official British contact with the press had been effectively handled by press spokesman Nic Fenn. Clearly something was up. In the preceding days newspapers had speculated that Carrington was preparing to proceed with a second-class solution. At the press conference Carrington gave substance to the rumors, which had been leaking out of Whitehall, without actually confirming them. He announced that he was entering into bilateral talks with the Muzorewa delegation on implementation of the constitutional proposals and elections. The message was clear: the train was leaving the station, and the Patriotic Front had scant time to jump aboard. The press reacted predictably, with the *Guardian* headlining, "Carrington Decides to Go Ahead without PF—Britain Pushes for Deal with Muzorewa."[20]

The hard British line brought a sharp response from Commonwealth Secretary General Ramphal, his first public criticism of the foreign secretary. "It would be a mistake," he said, "to assume Commonwealth support for any procedure at variance with [the Lusaka agreement]."[21] A Carrington-Ramphal meeting on October 16 resulted in a softening of tone on both sides, with the latter returning to his by-now customary role of seeking common ground. Although Ramphal's interventions were generally helpful, the perceived sense of crisis in the conference brought another visitor to London who was potentially more troublesome, South African Foreign Minister Roloef "Pik" Botha. For the British his hurried visit was like that of a boring rich uncle, unwelcomed but warranting attention. Botha's intentions were probably to encourage the Muzorewa delegation to hang on while pushing the British to take the plunge into the second-class solution. On October 17 he met with Mrs. Thatcher and Carrington and left apparently satisfied with what he heard. For the British his visit was a success in that he, at the least, seemed to cause no major difficulties. Moreover, his presence, when coupled with the arrival in London of Lt. General Peter Walls, Rhodesia's chief military figure, and the return to the conference of Ian Smith, lent credibility to Carrington's public statement that he was beginning implementation discussions with the Muzorewa delegation, and to the speculation that the British were moving toward a separate deal with the bishop.

Perhaps the most significant involvement of a supporting actor was that of U.S. Ambassador Kingman Brewster. Responding to separate requests from Ramphal and Carrington, President Carter authorized Brewster to convey to the British, the Front Line states, and the Patriotic Front a pledge of U.S. assistance should Lancaster House result in a success. The wording of the U.S. commitment was convoluted and cautious, reflecting the Carter administration's concern that it might face congressional criticism for participating in a "buy out" of white landlords or for opening the U.S. treasury to land-hungry peasants.[22] The U.S. offer of October 15 did not significantly add substance to Carrington's pledge of a few days before, but it did present Nkomo and Mugabe with a face-saving way out of the impasse, which, finally, on October 18 they took. Nkomo told the press that "if the U.S. had not stepped in it would have been very difficult to move on this question."[23] In a statement to a plenary of that date, he noted that the British and U.S. assurances on land issues

> go a long way in allaying the great concern we have over the whole land question. . . . We are now able to say that if we are satisfied beyond doubt about the vital issues of the transitional arrangements, there will be no need to revert to discussion on the constitution, including those issues on which we reserve our position.[24]

From the fourteen objections to the British constitutional proposals noted in its October 8 statement, the Patriotic Front had been whittled down to focusing on the land issue and had been compelled to accept undoubtedly sincere, but still vague, promises of assistance.

Another factor that may have played a role in their acceptance was the knowledge, now leaked to the press and
discussed in general terms with Nkomo and Mugabe, of British
intentions to take an active role in the transitional period,
thus limiting, to some degree, their continued concern about
Rhodesian regime control during the interim period. Neither
the Patriotic Front nor the Muzorewa delegation would be
pleased with the U.K. transitional plans as initially outlined.
The British would find themselves making deeper commitments
than they had originally intended as the conference moved
into its second act.

5
The Transition

It had taken the Lancaster House Conference six weeks to hammer out an agreement on the constitution. The next phase, to determine the arrangements for the transitional period during which elections would be held, was universally viewed to be even more fraught with difficulty.

On September 18, that is, after the Patriotic Front had accepted Carrington's agenda and before the Muzorewa delegation formally agreed to the broad British constitutional principles, Nkomo and Mugabe had issued their own plan for the transition. Their proposals suggested the creation of an eight-man transitional governing council, composed of four Patriotic Front members and four British/Salisbury regime representatives, with a British chairman. Subsidiary commissions with similar compositions would direct the army, police, public service, and judiciary. A United Nations force would keep the peace.[1]

This plan was designed to neutralize the de jure and de facto power of the Rhodesian regime while aggrandizing for the Patriotic Front the lion's share of interim authority. As an opening bid, it was consistent with the negotiating history since 1976, which had focused on power sharing during the transition. The Muzorewa delegation did not present a transitional plan of its own. Even after his grudging acceptance of new elections, the bishop maintained that no further alterations to the Salisbury regime would be necessary: his government would organize the elections and otherwise administer Rhodesia during the transition.

The British had given serious thought to the transitional period. It was clear to Carrington that Britain would have to exercise direct control in Rhodesia during the interim. Complicated power-sharing arrangements that had formed the basis of the Kissinger and Anglo-American plans were impractical, unlikely to be agreed upon, and probably unworkable in practice. The British governor would have to be precisely that, a governor with full executive and legislative authority. This would mean that the transition would be a period of intense political exposure for Britain. To limit its vulnerability, Carrington determined from the outset that the transition would have to be of the shortest possible duration. Its brevity would not allow for any restructuring of the governmental apparatus already in place.[2] This last point was also consistent with the basic British assumption that it must formulate its proposals to convince those possessing power that the new arrangements would not be so inimical to their current status as to make them unacceptable. Significantly, not envisaged in the initial U.K. planning was any role for UN officials or a peacekeeping force, both important elements of the Anglo-American plan and both particularly objectionable to the Salisbury regime.

After several days of informal discussions, Carrington issued on October 22 a thirteen-paragraph proposal for the transitional period. The document was similar both in style (relatively undetailed) and purpose (to focus conversation on a British draft) to the initial constitutional proposals. The U.K. plan centered on the assumption of direct British control through the appointment of a governor. He would have overall control of the elections, assisted by an election commissioner and "witnessed" by Commonwealth representatives. An advisory electoral council composed of representatives of each party would assist the commissioner. The commanders of the opposing armies would be responsible to the governor for maintaining a ceasefire. Law and order would be the responsibility of the established police force, also responsible to the governor. Key to the U.K. plan was its explicit recognition that the transition would bring no changes to the Salisbury administration. Such change, the British argued, would "prejudice or preempt the freedom of choice of the people of Zimbabwe."[3] After elections, the British said, a new government could restructure the administration, consistent with the constitution, but during the transition the governor would utilize the established administrative machine.

The proposals struck sensitive nerves in both the Salisbury and Patriotic Front camps. Though not specifically stated in the document, the advent of a British governor with full legislative and executive power could only mean the relinquishment of power by the bishop and his cabinet, a bitter pill for a newly elected government. The plan was equally objectionable to the delegation of Mugabe and Nkomo—no power sharing, no UN presence, no peacekeeping force. Most galling was the reliance on the established civil service for administration and the police—a partially paramilitary force integrated into the joint operation command of the Rhodesian military—for the maintenance of peace. This was designed, they claimed, to "clearly exclude the Patriotic Front forces from the security forces at the disposal of the Governor. The British Government is saying quite clearly that it is the Patriotic Front forces who would prejudice or preempt the freedom of choice of the people of Zimbabwe but that the Regime's forces would not."[4] Adding injury to insult was a massive Rhodesian Security Force raid into Zambia, destroying road and rail bridges and ZAPU military installations, launched as Carrington was presenting his proposals by those same security forces upon which the British, in the view of Mugabe and Nkomo, were intent on basing security for the transitional period. The Patriotic Front also objected to Carrington's timetable, not outlined in the proposals, but otherwise made clear by British officials who argued that two months would be a sufficiently long period. The Patriotic Front asserted that two to three times that amount of time would be necessary.

The days following the presentation of the proposals witnessed difficult British negotiating sessions with the Salisbury delegation. Obtaining from Muzorewa his agreement to step aside and transfer power to a British governor proved to be the single most difficult task confronting Carrington. Clearly, the most important individual in the process of convincing Muzorewa was General Walls, then in London and directly exerting his influence on Muzorewa and others.[5] Walls worked closely with two other Rhodesian security professionals, Ken Flower, head of the Central Intelligence Organization, and Harold Hawkins, former chief of the Air Force and then Rhodesia's representative in Pretoria. The most appealing element of the British plan to these professionals was the same heavy reliance on the established civil and security apparatus that so angered the Patriotic Front. On October 28 Bishop Muzorewa accepted the initial U.K. transitional proposals. He had taken almost the exact length of time as he had to signal acceptance of the first constitutional plan. Carrington publicly praised the bishop for his "statesman-like" act, implicitly contrasting him with Nkomo and Mugabe, who were putting up stiff resistance.[6]

The Patriotic Front's positions were strongly supported by the Front Line states and the Commonwealth secretariat, which had already raised their concerns with the U.S. government. At the same time Ambassador Brewster had been authorized to make the offer of U.S. financial assistance for rural development after independence, he had also been instructed by Washington to tell Lord Carrington that the Carter administration believed that a four- to six-month transitional period would be necessary and that additional measures to the ones then contemplated by the British would be necessary to meet Patriotic Front concerns about impartiality during the rule of the British governor. In late October Secretary Vance sent his Policy Planning Director, Tony Lake, to London to make the same points in person to the British Lancaster House team.

Once again Ramphal, with U.S. support and assisted by Jamaican Prime Minister Manley, who was briefly visiting London, conducted a mediating effort between the Patriotic Front and the British, which, by November 1, resulted in some indication of British flexibility on certain issues. On that date Carrington stated that he would be prepared to extend the time allocated for the transitional period for two to three weeks more, if necessary, to allow for a ceasefire to take hold.

At a plenary on the following day Carrington presented an amplified forty-one point transitional plan.[7] Several bows were made in the Patriotic Front's direction: British police officers would assist the governor in supervising the local constabulary; all political prisoners would have their cases reviewed; a start would be made on the repatriation of refugees; political parties would be unbanned; and all facilities would be granted to Commonwealth observers to allow them an unimpeded view of the elections. Despite the modifications, the essentials of the British plan remained, and the major elements of the Patriotic Front's position continued to be disregarded. At a press conference on November 3, Nkomo and Mugabe repeated their basic objections, but were careful not to reject the plan outright.

Two days later Muzorewa formally accepted the forty-one points. His motivation was similar to his earlier acceptances: to put the Patriotic Front on the defensive, perhaps force their walkout from Lancaster House; and to position the Salisbury regime where it might obtain the best possible deal from the British on the crucial issues of recognition and sanctions. The latter assumed even greater significance as the date for the yearly vote on extending sanctions approached in Parliament. The press had published reports earlier of the British option on sanctions that was used to discourage Smith in the first days of the conference, that is, not seeking an extension but retaining the 90 percent of the sanctions covered by other legislation or authority. Nevertheless, it was still not clear what path Carrington would pursue. He had been vague at the Tory Congress, and the press had speculated that he might seek not a year's renewal, but rather a monthly extension. By accepting the transitional plan, the Muzorewa delegation positioned itself to encourage back-bench Tory support for a revolt should Carrington seek either the yearly or the monthly renewal. At the very least, the Salisbury team realized that their acceptance would make it difficult for Carrington to take what would be seen as an insulting step to Britain's cooperative conference partner by seeking a parliamentary resolution for extension of any length. Salisbury wanted all sanctions lifted, but some on Muzorewa's delegation realized that this would be difficult for Britain, fomenting serious strain in its international relations.

The line Carrington adopted was a masterful interplay of parliamentary maneuvering and conference pressure tactics. He was able to placate the Muzorewa team, avoid a showdown with his own party's right wing, and, at the same time, take symbolic and political steps to reinforce with the Patriotic Front that the second-class solution was still very much alive. On November 7 Sir Ian Gilmour told the House of Commons that Mrs. Thatcher's government would not seek to renew the sanctions legislation, noting, however, that most would remain in force. More significantly, he introduced legislation that would enable the government to implement portions of the independence constitution, to appoint a governor for Rhodesia and to hold elections. The legislation was a sweet gift for Muzorewa. It indicated that Britain intended to move ahead swiftly, with or without the Patriotic Front, and that the lifting of sanctions could not be far behind. Though Gilmour specifically denied Labour charges that the bill was designed to promote a separate agreement with the bishop, he increased suspicion when he refused to guarantee that its provisions would not be implemented until an agreement with all of the parties had been achieved.

The introduction of the enabling legislation angered the Patriotic Front, inflamed their suspicions of the British, and heightened what one press account termed the conference's "atmosphere of crisis."[8] Into this tense scene rushed President Kaunda, reeling from continued Rhodesian military raids into his country and Muzorewa's November 5 decision to cease maize exports to Zambia. Over the weekend of November 10–11 Kaunda met on three occasions with Mrs. Thatcher and Carrington and several times with Nkomo and Mugabe, as well as with Ramphal. Journalists sympathetic to Kaunda credited him with "thawing the talks,"[9] though it seems clear that modifications to the British proposals were in train prior to his arrival in London. On Friday, Britain had announced at a plenary session that it would be prepared to contribute men to a Commonwealth military monitoring force. News from Australia and New Zealand confirmed that those two nations had been asked by Carrington to commit troops. A Commonwealth force of several hundred men, clearly described as "monitors," not peacekeepers, was not the multithousand UN army Nkomo and Mugabe had been demanding. But it was another indication of British sensitivity to the Patriotic Front's concern, echoed by Ramphal, Kaunda, and others, including senior officials of the Carter administration, that some military element, apart from the Rhodesian police and the opposing armies, would be necessary to ensure impartiality.

The decision to commit British troops, which had not been a part of Carrington's preconference planning, was not made lightly and represented a significant gamble on the part of Mrs. Thatcher and her foreign secretary. But they apparently determined that sending a small group of men, carrying little more than lighted candles into a powder keg situation, was worth the risk. In one stroke it undercut the Patriotic Front's demands for UN participation while involving the Commonwealth in a manner likely to reinforce its commitment to a peaceful transition and its willingness to use its influence on Nkomo and Mugabe to moderate their positions.

But the monitoring force did not satisfy all of the Patriotic Front's objections. The tense debate continued in plenary sessions and in private meetings. Meanwhile, the enabling legislation moved through Parliament, giving Carrington and Gilmour the opportunity to air the threat of a second-class solution once again outside of the conference hall in which it might have appeared too bullying in nature. In part, their parliamentary comments aimed to mollify the Tory right wing, as well as affect the Patriotic Front (discouragingly) and the Salisbury delegation (encouragingly). Carrington told the House of Lords that he still wanted a settlement acceptable to all, "but, if this is not attainable we cannot allow the best to become the enemy of the good. With an agreement already reached on genuine majority rule, and an end to the rebellion against Britain's authority, there can be no turning back."[10]

While threatening to move ahead without all of the passengers, Carrington was keeping the train at the station. In three days of intensive bilateral negotiations with the Patriotic Front he remained firm on the essentials but offered some concessions. Meanwhile, Whitehall officials leaked stories to the press indicating an imminent split within the Patriotic Front if Mugabe did not go along with Nkomo's more moderate position. The British may have based the self-serving rumors on their own hopes or on intelligence gathered from members of that delegation, including Nkomo himself, and from the bugging of the delegation's telephones and hotel rooms.[11] Whatever the source or veracity of the information, the British intent was clearly designed to increase pressure on Mugabe. At a November 14 meeting Carrington reportedly told the Patriotic Front's leaders that he had to have their answer before the following day's cabinet meeting, when he expected to come under pressure to move ahead with the bishop's agreement only.[12] But by the time of that warning, the outlines of a compromise had emerged.

As in the first act, the Patriotic Front's many objections to the U.K.'s proposals had been whittled away in negotiation. Some had been rejected, others had been addressed in British redrafts, often obliquely so as not to arouse the Muzorewa delegation. Still others had been defused by private British explanations that elements Nkomo and Mugabe found objectionable on paper would be inoffensive in practice. What remained now was their insistence that their armies be treated by the governor in an identical fashion as the Rhodesian security forces during the transitional period. This was more than a question of pride. It related to their concern that if law and order deteriorated to the point where it could not be restored by the police alone, the governor would assume their separate or joint culpability and rely solely on the Rhodesian military for help. Those forces would, in turn, use their operational freedom to the Patriotic Front's disadvantage.

The actual and symbolic content of the issue was no less important to the Rhodesian military. In the view of the white officer corps, the granting of equal status to the guerrilla armies would prejudge the election process, giving de facto and de jure power to entities that, it could be argued, had not electorally proven themselves to have popular support. More importantly, the Salisbury delegation as a whole feared that equal status would be misused, become a pretext for free-roaming depredations of the local population, and, not incidentally, panic the white community. The British were in a bind. At its irreducible minimum, the Patriotic Front's demand seemed justified to the Front Line states and understandable to the United States, all of which made their views known to Carrington.

Finally, on November 15 the British were able to reach an agreement with the Patriotic Front that was salable to the Rhodesian military. The United Kingdom would agree to pay, house, and feed the guerrillas during the transitional period. More importantly, from the Patriotic Front's point of view, a handful of words was added to one of Carrington's forty-one points, giving their forces additional responsibilities, or more precisely, a responsibility that was already implicit elsewhere. The thirteenth point had ended with:

> Executive authority will be vested in the Governor and all public authorities in Rhodesia, including the civil service, the police and defence forces, will be required to comply with the Governor's direction.

Carrington agreed to add as the new final sentence: "The Patriotic Front's forces will also be required to comply with the directions of the Governor."[13]

In accepting the compromise language, Nkomo and Mugabe may have been influenced by British discussion of elements of the ceasefire provisions, the next phase of negotiation, which could be interpreted as favorable to them. Similar revelation of transitional period thinking had helped bring the constitutional discussions to an end. As one British official told the press, "We let them have a glimpse of the goodies awaiting them."[14]

Actually obtaining an amendment to the forty-one points, albeit a minor one, was something of a victory for Mugabe and Nkomo. They had been unable to convince the British to delete or add a word to the final constitutional proposals. Nevertheless, their claims of victory rang hollow. The essentials of the British plan remained unchanged, and the same British tactics used in the first act—utilization of early Muzorewa acquiescence and threats of pursuing a second-class solution—proved too much for the Patriotic Front once again. Carrington's insistence on the step-by-step approach was bearing fruit and putting Nkomo and Mugabe in an increasingly disadvantageous bargaining position. As a high-level British official (perhaps Carrington himself) told the *New York Times:*

> The Front is committed to so much now—the constitution, the transition—that they would be hard put to sell an election boycott to the rest of the world. If we carry out what we have agreed to in good faith, even if the Front walks out at the last minute, we will have a fair part of Africa and much of the world with us.[15]

Flushed with victory, the British moved on to the last phase, likely to be even more difficult, for it would deal with the literally life-and-death issues of the ceasefire.

6
The Ceasefire

The inherent tensions of the ceasefire issue were, if anything, greater than those of the constitution and the transitional period, but other factors gave the third act, so similar in some ways to the two preceding it, a special style. Likelihood for success was increased precisely because it was the third and final act. It had been preceded by almost eleven weeks of hard bargaining in which the currency of threat, promise, counterthreat, and associated theatrics had been cheapened by heavy usage. Momentum had been achieved and was now a force to reckon with. The momentum's momentum, however, was not so great as to assure success.

Post-conference accounts which assert that "an air of inevitability hung over Lancaster House" clearly overstate the case.[1] Even after close to three months of negotiation, important actors, including both Carrington and Mugabe, doubted the prospects of achieving a settlement that would involve all of the parties. The momentum needed to be helped along, thus prompting a British decision, acquiesced in by the Patriotic Front, to bifurcate the ceasefire negotiations after they had already begun. Successfully concluding the first phase maintained the necessary forward progress that, nevertheless, still could have dissipated during the lengthy discussions of the second phase.

The British felt that they needed all the help they could get to maintain pressure on the Patriotic Front. Their sensitivity on this issue prompted one of the touchier episodes in Anglo-American relations during the Lancaster House period. Cyrus

Vance describes the incident as a tactical difference that became "a rather serious irritant."[2] On November 14, President Carter, responding to a legislative dictate that had been part of a previous compromise to maintain sanctions, reported to Congress that he would be prepared to lift sanctions only when a British governor arrived in Salisbury and the process leading to impartial elections had begun. Carrington had wanted the Carter administration to state that U.S. adherence to sanctions would end simply when the governor took up his position. Mention of the electoral process, which to the United States had been the keystone of the settlement efforts, was interpreted by the British as increasing the Patriotic Front's leverage, not Carrington's.

Vance writes that the U.S. decision was warmly received by the Patriotic Front and the Front Line states, which were still worried that if sanctions were terminated before elections, "the Rhodesian authorities would find excuses to postpone or avoid them." What Vance does not say, and what undoubtedly added to the British bitterness, was that the African suspicion also extended to Her Majesty's Government and that President Carter seemed to be implicitly bolstering and accepting that lack of trust.

The British were not, however, going to allow the flap with the United States to deter their decision to force the pace of the conference. On the morning following the Patriotic Front's agreement to the transitional proposals, Carrington introduced for discussion a ten-point ceasefire plan. It called for a ceasefire to commence within seven to ten days of the completion of a Lancaster House agreement. By that time all cross-border infiltration of Patriotic Front troops would have ceased and those guerrillas within the country would gather with their arms at designated assembly points. In maintaining the ceasefire the governor would be assisted by the Commonwealth monitoring force and a committee upon which the commanders of the three armies would sit. The Salisbury delegation's response to the proposals was mild, reflecting a month of direct British negotiation with General Walls. Meanwhile, the Rhodesian Security Force was in the midst of its third major raid into Zambia since the conference's outset. As in past attacks, the motives were probably mixed and perhaps contradictory. By destroying road and rail bridges on all Zambian highways leading to Tanzania and Malawi, thus making that country even more dependent upon Rhodesia for transport, the Rhodesians were clearly seeking to increase their leverage on Kaunda. The raids were also designed to hamper ZAPU troop movements. An unintended consequence of the raids may have been to increase the Zambian leader's desire to push Nkomo and Mugabe to participate in the very type of settlement the Rhodesian military still did not want.

The Patriotic Front's response was equally predictable and followed the pattern established in earlier negotiations. Three days after Carrington had presented his plan, Nkomo and Mugabe offered their own. It called for a Commonwealth "peacekeeping" (not monitoring) force of several thousand, the disbandment of certain Rhodesian military units, and a ceasefire supervisory commission (not an advisory committee). The Patriotic Front would sit on the commission, which would have the authority to decide when the ceasefire had become sufficiently effective to allow elections to be held, a period likely to last several months, not the eight to ten weeks the British envisaged for the entire transitional period. The counterplan also called for the surrender of the "vast private armoury" of weapons in the hands of Rhodesia's white civilians.[3] Following past practice, Carrington refused to accept the Patriotic Front's position paper as a basis for discussion, preferring to use his own.

Interestingly, at this point in the conference Nkomo and Mugabe made a concerted effort to negotiate directly with General Walls rather than follow the now standard procedures of bilateral talks with the British and plenary sessions. They viewed Walls as the most powerful Rhodesian player and felt that their own military men, notably Tongogara, might be able to establish a decent working and negotiating relationship with him. Whether Walls or the British felt this would be useful is unclear, for the offer to negotiate was rejected by Muzorewa, now returned to Salisbury to begin campaigning.

The bishop was acting out of pique. Nkomo and Mugabe had refused his suggestion to meet with him earlier in the conference and he was not now prepared to grant their request to meet with the commander of his armed forces.

On November 24 Carrington released his amplified ceasefire proposals. They expanded his November 16 ten-point plan with such details as where the Commonwealth monitors would be stationed, i.e., at each assembly point of Patriotic Front forces and at every Rhodesian command center down to the company base level. In presenting the document Carrington asked for an answer within two days. Dutifully, on Monday, November 26, the Muzorewa delegation formally accepted the expanded proposals. For their part Nkomo and Mugabe flew off to Dar es Salaam for a meeting with the Front Line presidents. Nkomo told a departure press conference that Carrington was guilty of "arrogance tinged with racism" and that the foreign secretary "can go to hell."[4] Despite the harsh words, there was no indication that the Patriotic Front leaders were walking out of the conference. The Dar es Salaam meeting discussed tactics. The Front Line presidents discouraged moves that would cause Lancaster House failure, but agreed to press on the British the need to withdraw Rhodesian troops from border areas where they could attack the guerrillas moving to the designated assembly points. Upon their return from Tanzania, Nkomo and Mugabe resumed negotiations with the British. The press picked up some signs of increased flexibility on both sides.

Into this environment intruded South Africa's foreign minister, who arrived in London for his second visit of the conference period. As on the previous occasion, Botha admonished the British for being too soft on the Patriotic Front and urged Carrington to put the second-class solution into play. Nevetheless, as in his first visit, he accepted British explanations of what was happening and left London satisfied that he had been sufficiently consulted, always a principal goal of the status-starved South African government.

Botha may have also warned the British about a statement that his Prime Minister, P. W. Botha, was soon to make. At a dinner honoring the retiring South African Air Force chief on November 30, Botha would reveal that South African forces had been operating within Zimbabwe "for some time," protecting transport routes to South Africa and making sure that anti-Pretoria "terrorists" did not take advantage of the Rhodesian war to attack southward.[5] The prime minister's statement was to become the first public acknowledgment that South African regulars were fighting in Rhodesia since the withdrawal of 2,000 Pretoria regime paramilitary police from Rhodesia in 1975 as part of an ill-fated South African detente effort with black Africa. While shocking to some, including many in the South African public who had been repeatedly assured that their troops were not involved to the north, P. W. Botha's statement would only confirm longstanding Patriotic Front suspicions and propaganda. Thereafter, the removal of South African troops from Rhodesia as part of the ceasefire agreement was a principal Patriotic Front demand.

As had happened previously, the longer the negotiations continued, the fewer the sticking issues became. In addition to the objectionable South African presence, the Patriotic Front opposed the size and compositon of the Commonwealth force, the perceived inequality of treatment of its own forces and those of the Rhodesian regime, and the length of the ceasefire period. Nkomo and Mugabe suggested several Third World Commonwealth members to augment the troops now expected from Britain, Australia, New Zealand, Kenya, and Fiji. Carrington, urged by Ramphal and U.S. officials, had already agreed to increase the force's size to over 1,000. But he would accept no additional contributor nations. Nor could Carrington accept the Patriotic Front's demand for a two-month ceasefire period to precede another multimonth period in which political campaigning and elections would take place. The British were adamant on keeping the transitional period as brief as possible in order to minimize the duration of their political vulnerability. At this point, General Tongogara, who the British found to be both more reasonable and militarily savvy than his political boss, Robert Mugabe, acknowledged to FCO officials that he could have orders delivered from

Mozambique to any point within Zimbabwe by runners within four days.

Tongogara's candor undercut the Patriotic Front's position that it would take weeks for their men in the bush to get the ceasefire word. This paved the way for the ultimate agreement on the issue—a seven-day assembly period to begin one week after the termination of the conference, a total of two weeks for the ceasefire to take effect—much closer to the initial British proposal of no more than ten days than to the Patriotic Front's demand for two months. The issue of force treatment parity, or "reciprocal disengagement" as it was termed by the Patriotic Front, was not as easily resolved and would provide the conference's final drama.

Despite Tongogara's pragmatic intervention, negotiations limped into the first week of December. Patriotic Front spokesmen were relatively upbeat. "A settlement is in sight and it's only a stupid person who would jeopardize it," ZAPU's press liaison told the *Observer*.[6] But Carrington was becoming increasingly impatient. On Monday, December 3, the acting chairman of the Salisbury delegation, Mundawarara, told Carrington that his team was packing up and leaving for Rhodesia in protest over Britain's continued kid glove treatment of the Patriotic Front. Carrington talked him out of leaving. But that night, perhaps engaging in a bit of theatrical hyperbole, the British foreign secretary told the press that "I do not despair of reaching an agreement, but I am as close to despair as I have been in the whole three months of negotiations."[7]

Not one to sit idly by and bemoan his fate, Carrington initiated another series of extra-conference maneuvers to pressure the Patriotic Front. On December 3, the necesary legislation through Parliament, he obtained an Order in Council from the cabinet giving Mrs. Thatcher the authority to select an individual to serve as governor of the colony of Southern Rhodesia. He also introduced for discussion another Order in Council, which would give the British government the legal authority to promulgate the new constitution. Carrington was sending a message designed to placate the Muzorewa delegation and spur the Patriotic Front to agreement. The threat of the second-class solution was being waved once again. This was more than a simple negotiating tactic. An imminent and irrevocable move toward implementation of an agreement with Muzorewa alone could not be ruled out.

At the same time, however, Carrington continued intensive bilateral negotiations with the Patriotic Front, again with the active involvement of Commonwealth Secretary General Ramphal. Finally, late on the evening of December 4, partial accord was reached. In the presence of Nkomo and Mugabe, Ramphal telephoned Tanzanian President Nyerere and convinced him that the agreement would not put the Patriotic Front at risk.[8] The keystone of the compromise announced by Carrington in the following morning's plenary session was an agreement to put off resolution of the reciprocal disengagement issue while accepting the other elements of the U.K. ceasefire plan. Additionally, Carrington stated at the plenary that "there will be no external involvement in Rhodesia under the British Governor. The position has been made clear to all the governments concerned, including South Africa."[9] It was his first public mention of the Pretoria regime in this regard. The foreign secretary also stated that measures would be undertaken to insure the neutralization of the Rhodesian Air Force. Taken together, these moves were enough to enable the Patriotic Front to enter into a partial ceasefire agreement.

In a conference characterized by continued Salisbury cooperation with the British, the Patriotic Front's decision to defer discussion on the assembly points and related issues and to participate in what, in reality, was a "hyped up" announcement of an incomplete accord, was a signal act of cooperation. It was designed to maintain momentum at a time when the Muzorewa team was threatening to jump off the negotiating wagon and Carrington was coming perilously close to a second-class solution. As if to twit his colleagues, Ian Smith, now in Salisbury, said, "As I have predicted, the British Government would never have gone ahead without the Patriotic Front. They have been bluffing all along the line."[10] Carrington pronounced himself "enormously encouraged" and told the

London press that only a few more days of negotiation would be necessary to wrap up the conference.[11]

The euphoria was short-lived. The conference was to enter another period of prolonged bickering (eleven days) that would witness Carrington's most daring gamble. Several days of bilateral negotiations yielded little progress. Meanwhile, over the weekend of December 8–9 the Rhodesian forces staged major raids into both Zambia and Mozambique, destroying roads, bridges, and guerrilla camps (often indistinguishable from refugee sites) in what may have been an effort to get their last licks in. The British wisely postponed a plenary for December 10, which would have been marked by Patriotic Front vituperation, reflecting their real anger over the attacks. But on the next day, Carrington took dramatic action to force the conference to its termination.

Several days earlier Lord Christopher Soames had been named as governor of Rhodesia. To some his appointment presented "confirmation that the Government believes a settlement is in the bag."[12] Conservative party leader in the House of Lords, former ambassador to Paris, member of the privy council, holder of other important positions, and Winston Churchill's son-in-law, Soames, though not a confidante of Mrs. Thatcher, was neverthless a political personage of some standing. His nomination reaffirmed Britain's commitment to see the Rhodesian settlement and transition through to a successful completion. On December 11 Carrington announced that Soames would leave immediately to assume his position.[13] On the same day the Rhodesian parliament unanimously voted to dissolve itself and accept British authority. Sanctions were lifted as Soames landed in Salisbury. The decision to send Soames was but another British action designed to convince both delegations, but, in this case especially the Patriotic Front, that a settlement, if necessary without their participation, was imminent and inevitable. Another factor in Carrington's decision may have been the desire to impose control over the Rhodesian Security Force and to end the cross-border operations that were threatening conference progress.

While waving the stick of Soames's departure, Carrington also offered Nkomo and Mugabe some carrots. In a plenary meeting on the day Soames left London, Carrington noted that the assembly points chosen for the Patriotic Front's fighters were near their operational areas and purposely distant from Rhodesian army bases. The monitoring force, now increased to 1,200 men in response to Front Line requests, would be in the assembly points providing symbolic protection. Moreover, to start the ceasefire, the Rhodesian forces would take the first step, "moving into the close vicinity of their bases to permit the Patriotic Front to assemble their forces."[14] He made it clear, however, that the success of the ceasefire period depended on the Patriotic Front's performance.

> If the Patriotic Front forces remain in the field or continue to be introduced from outside the country, those forces will be unlawful. If, however, all Patriotic Front forces inside Rhodesia assemble with their arms and there is no further movement by externally based Patriotic Front forces into Rhodesia, there would be no need in those circumstances for the Governor to ask the Rhodesian forces to deploy from their company bases . . . There could, in these circumstances, be no danger to their [the Patriotic Front's] security. I am conscious of the concerns expressed by the Patriotic Front that their assembly places should not be in close proximity to Rhodesian bases and that they should not be "encircled." There has never been any question of the Patriotic Front being encircled. They will be under the authority of their own commanders and other forces will be in close proximity to them.[15]

Carrington's December 11 statement spoke to the Patriotic Front's concern about becoming sitting ducks for the Rhodesian army, but there were other reasons Nkomo and Mugabe objected to the plan to assemble their troops into fifteen assembly points, and these were largely political in nature. Why should the Rhodesian security forces, they asked, be allowed to maintain more than ninety visible bases while their own forces would be relegated to a much smaller number in remote areas? They argued that the local population would view this as a victory for Muzorewa and lose its confidence in the Patriotic Front's fighting and political capacity. Nkomo and Mugabe were angered by the dispatch of Soames and further infuriated by Carrington's December 14 ultimatum to agree by the following day, when he would end all plenary sessions and depart with Mrs. Thatcher for an official visit to the United States. ZANU spokesman Eddison Zvobgo told the press that conference failure held no fear for the Patriotic Front: it would simply mean "an all-out war with a British Governor in charge of Rhodesia."[16] Notwithstanding the public bravado, behind the scenes both the British and the Patriotic Front were searching for a way out.

At the scheduled plenary on December 15 Carrington bent further on the assembly points, noting that

> if the Patriotic Front forces at present in Rhodesia assemble with their arms and equipment in numbers greater than can be dealt with at the assembly places designated in the ceasefire agreement, the Governor will assess the need for additional sites in relation to the successful accomplishment of the assembly process by the Patriotic Front forces and in relation to the disposition of their forces.[17]

In other words, Carrington held out the possibility of additional assembly points if the number of troops the Patriotic Front was now claiming would assemble, over 30,000, and which both the British and Walls thought highly exaggerated, did in fact assemble.

Having made the statement, Carrington initialed the final settlement accord with Mundawarara and flew off to Washington. The United States had been broadly and publicly supportive of the British during the conference. Ambassador Brewster and his Africa watcher, E. "Gib" Lanpher, had played active, behind-the-scenes roles, facilitating communication between London and Washington and among the London actors. Nevertheless, many policymakers in Washington shared Commonwealth and African suspicion that Carrington was pushing the Patriotic Front too hard and perhaps really did not want a first-class solution. The British had taken pains to keep the U.S. State Department informed of Lancaster House developments, but had not appreciated the periodic advice on conference management emanating from Washington. President Carter was anxious to welcome Mrs. Thatcher, who had been supportive in the then six-week-old Iran hostage crisis, as handsomely as possible.

Several days before leaving for Washington Carrington had asked that the United States join with Britain in lifting sanctions on the day of Soames's arrival in Salisbury. According to Vance:

> We told the British that the dispatch of the governor general did not meet our requirements for terminating sanctions. With some grumbling about our "self-created difficulties," the British arranged that upon Soames's arrival the election process would begin. An election commission was established, voter registration procedures were promulgated, and assurances were given that all parties could campaign freely. With these steps taken, the President issued an executive order on December 16 terminating sanctions against Rhodesia.[18]

The U.S. position was an understandable one that sought to use the leverage of the moment to obtain previously established goals. President Carter's November 14 statement to Congress that sanctions would be lifted only when the British governor arrived *and* the electoral process had begun added further urgency to the U.S. request.

However, the British perspective on the matter was, not surprisingly, quite different. As they saw it, the prime minister and foreign secretary of the United States' closest ally were preparing to visit Washington immediately after taking the most audacious gamble of their governance. The response to their request for U.S. assistance resembled a lawyer's brief of conditions necessary and precedent to completion of a contract between two not very trusting business associates. The British were joined in their anger by the U.S. ambassador in London, Kingman Brewster, who, according to embassy officials, threatened to resign if the British request to lift sanctions was not acceded to prior to Mrs. Thatcher's and Carrington's arrival in the United States. His threat, more than the British actions outlined by Vance, tipped the scales. An Executive Order rescinding U.S. compliance with sanctions was issued before the visiting British touched down in New York.

The president's move received wide congressional support, except from several House members who had led the fight on the president's behalf to maintain sanctions. They were miffed by his lack of consultation with them and concerned that removal of sanctions was unwise in the absence of a comprehensive settlement in London, approved by the UN.[19]

While Carrington and Thatcher visited the United States, negotiations with the Patriotic Front continued in London. The foreign secretary's December 15 statement had not fully satisfied Nkomo and Mugabe. As in the protracted debate on the transitional period, it became important for them to demonstrate their own negotiating strength by forcing the British to actually amend the proposals. Over the weekend of December 15–16 the United Kingdom formally offered a sixteenth assembly point to be located somewhere in the center of Rhodesia. Sir Antony Duff, already in Salisbury as Soames's Deputy Governor, had cleared the offer with General Walls in a difficult negotiating session on Saturday night,[20] which he was later to describe to confidantes as "the worst two hours in my life."

Nkomo was satisfied, but Mugabe still balked. His motivation was complex. He maintained his initial distrust of the British, though he had come to a grudging respect for Carrington. He was fearful of the designs of the Rhodesian military on the assembled guerrillas. More committed to revolutionary theory than Nkomo, he was also convinced that his forces would ultimately triumph militarily if an agreement was not reached. Finally, there may have been a more subtle reason causing Mugabe to pull back from a final agreement: he may have been afraid. As one author has written of a guerrilla leader who operated in a country far from Rhodesia nearly seventy years earlier, he was

a man obsessed with staying true . . . he could not betray a
promise for the life of him. But courage of one kind can hint
at cowardice of another; [he] was afraid—not for himself, but
of himself, of unwittingly betraying the trust his peers and their
people had invested in him.[21]

The ZANU leader decided that he would take his case to
the world. Over the weekend he contacted the U.S. embassy
to ask for visas for himself and aides to travel to New York,
where he would denounce the Lancaster House proceedings
at the very time Mrs. Thatcher and Carrington were visiting
the United States. The U.S. embassy deflected the applications,
pleading that no action could be taken until the start of the
business week on Monday.

Meanwhile, the British pulled out all the stops. The FCO
leaked to the press the information, accurate or not, that
Nkomo was about to sign the agreement without Mugabe.
They also encouraged General Tongogara to reinforce with
Mugabe his military acceptance of the plan. In fact, Tongogara
had been less concerned all along than the politicians about
the peripheral locations of the assembly points. Even though
he was aware that the sites had actually been selected by
General Walls, for him there was some advantage in having
his men near the borders in case they had to run for it if
the ceasefire fell apart. In those final days he influenced Mugabe
as did, and most crucially, President Machel of Mozambique.

Throughout the conference Mozambican representative Fernando Honwana had been the best informed and most active
of the Front Line observers. Now he enlisted the direct
assistance of his chief of state. On Machel's instructions,
Honwana told Mugabe to stop quibbling and to sign the
agreement: electoral victory was assured. In a very tough
statement, Honwana repeated to Mugabe Machel's message:
if he did not sign the agreement, he would be welcomed back
to Mozambique and given a beach villa where he could write
his memoirs. In other words, as far as Mozambique was
concerned, the war was over. The combination of Tongogara,
Nkomo, Machel, and conference momentum finally took its

toll. Mugabe joined Nkomo in agreeing to the sixteenth assembly point and on December 17, while Carrington was still in New York, the Patriotic Front's leaders initialed the agreement.[22]

Muzorewa, in a state of pique, delayed the formal signing for a few days, but by December 21 he had returned to London for the official signing ceremony on the one hundred and second day of the conference. Lancaster House was over, but the drama was not yet finished.

7
Denouement

Perhaps the most remarkable fact about the Lancaster House Conference was that the settlement achieved there actually accomplished what it set out to do, that is, provide for a peaceful transition to majority rule through free and fair elections. The impact of Lancaster House's three acts was felt in reverse order to their negotiating schedule.

After some rocky first days in which the expected flow of guerrillas into the assembly points was disappointingly low, the numbers entering picked up rapidly. By January 4, the end of the two-week ceasefire period, 17,000 had entered the camps and more straggled in later, until the numbers reached about 22,000—16,500 ZANU and 5,500 ZAPU.[1] Additional assembly points were not needed. Many guerrillas remained in Zambia and Mozambique. An estimated one-third of ZANU's forces within Rhodesia were ordered by their commanders not to enter the camps. They were told to bury their arms and meld in with the local population. Their presence in the countryside and charges of their intimidation of non–Patriotic Front supporters became a major issue in the electoral period. General Tongogara's death in a Mozambique road accident a few days after the signing of the Lancaster House agreement was a serious, but not deadly, blow to the ceasefire process.

The issue of South African troops was to bedevil the early days of the transition. Despite Carrington's apparent promise at Lancaster House, some of the troops were not withdrawn. The British had been told by General Walls and others that there were 500–1,000 South African army members integrated into Rhodesian Security Force units. What they perhaps had not been told was that South Africa had stationed three companies of troops in the vicinity of Beit Bridge on the Rhodesian side of the border. The troops were there to guard the vital rail and road link to South Africa, and to provide a convenient access point for more South African troops in case of massive chaos in Rhodesia. In order to placate Walls and the South African government, Soames was willing to finesse the issue, but when the continued presence of the troops became public knowledge in early January, the British were accused by the Patriotic Front and a good deal of world opinion, as expressed in a UN resolution, of hypocrisy or worse. After several weeks of intense private discussions, Pretoria publicly withdrew the three companies, although the South African "volunteers" remained integrated in Rhodesian Security Force units.[2]

The ceasefire was not perfect and came close to crashing down on several occasions as the Rhodesian military sought to punish ZANU troops they believed were violating the agreement's terms. However, throughout the transitional period the ceasefire held, often by the skin of its teeth, and only through effective management and negotiation by British civilian and military officials.

The tactics used effectively by Carrington at Lancaster House were echoed by Soames and his staff, which included Renwick, Duff, and other conference veterans. The difference in the British treatment of the contending parties was manifest. Operating on the assumption that it was necessary to keep those with power satisfied, Soames and his aides expended considerable effort in maintaining their relationship with Walls and the Rhodesian military. The British had to draw on an ever-decreasing store of goodwill to cajole, wheedle, and otherwise to persuade (often in bitter private combat) the Rhodesians to abide by the agreement and not respond to what Walls and others viewed as the Patriotic Front's provocations.

Some in the Rhodesian military were determined to foment a coup or force the Patriotic Front to pull out of the elections. An attempt on Mugabe's life failed and, though the assailants were never identified, elements of the Selous Scouts (the Rhodesian army's dirty tricks outfit) were the most likely suspects. Meanwhile, ZANU intimidation of voters in the countryside continued as did that by soldiers, in and out of mufti, loyal to Muzorewa, Nkomo, and other candidates.

To keep Walls and Muzorewa in line and also because the British themselves mistrusted the Patriotic Front, particularly Mugabe, Soames adopted a publicly more hostile posture toward the ZANU leader. Threats to disqualify all or some of Mugabe's candidates or to make null and void results from areas in which ZANU intimidation was reportedly rife filled much the same need for the British during the transition as the second-class solution had during Lancaster House. The tactic worked. British election observers in rural areas noted a decrease in intimidation in the three weeks preceding the elections and, as at Lancaster House, the threat was not implemented.

The elections were managed much in the manner the British had envisaged. White Rhodesian civil servants acted in a generally loyal and competent manner, with no small encouragement from Lord Soames's staff. Nearly one million more Africans cast their ballots in the last three days of February than had voted just eleven months before. The Commonwealth observer group, representatives of other organizations and governments, and hundreds of journalists judged the elections to be free and fair. They also determined that intimidation of voters had not played a significant role in the results.

Mugabe had rejected Nkomo's plea that the Patriotic Front run as a unified party. Contesting separately, Mugabe's ZANU captured fifty-seven of the eighty African seats. Nkomo's ZAPU took twenty in his home area of Matabeleland, and Bishop Muzorewa, the resounding victor of the April 1979 elections, whose confidence in his ability to win again had sustained him through Lancaster House and the transition, was able to retain only his seat and two others for his party. Other African parties were totally wiped out. Ian Smith's Rhodesian Front easily won all twenty seats in earlier scheduled voting for the country's white voters. The possibility of a coalition government of Muzorewa, Nkomo, and Smith had been highly touted by many Rhodesian whites and some British officials during the campaign. Soames and his closest staff had kept all options open pending the electoral results, but most probably would have favored a grand coalition that would have included Mugabe as well. But Mugabe's overwhelming victory saved Lord Soames, who was responsible for designating an individual to form the first government, from difficult decisions. He immediately asked the ZANU leader to become Zimbabwe's first prime minister. Mugabe, in turn, offered cabinet positions to Nkomo and some of his followers, though refraining from establishing an official coalition government, which, with its absolute majority, ZANU neither needed nor wanted.

Once the election results were known, Mugabe and Soames overcame whatever personal animosities remained between them and established a good working relationship. The British extended the period between elections and the formal granting of independence (to April 18, 1980), but not for as long as Mugabe had requested. Prince Charles represented the Crown at the independence ceremonies. In addition, Carrington and Mrs. Thatcher approved of a continuing British military presence in Zimbabwe to form a training and advisory team to ease the integration of the three former armies. The U.K. offered considerable aid to the new government and, with the United States, became Zimbabwe's major aid donor.

Taken together, ZANU's and ZAPU's seventy-seven seats would be sufficient to amend unentrenched provisions of the constitution they had objected to so virulently at Lancaster House. But, in fact, this has not happened and that document remains largely intact. Following independence a new dynamic has evolved in Zimbabwe in which the government has sought to demonstrate adherence to the Lancaster House agreement as a means of maintaining the support of the economically critical domestic white community and of those Western governments and international agencies that have supplied high levels of financial assistance.

The threats, both real and perceived, that the Mugabe government believes are posed to it by its opponents, including dissident Nkomo supporters and the South African government, have led to an unfortunate and heavy reliance on the security legislation and practices of the Smith regime. This has resulted in a chipping away at the spirit, if not necessarily the letter, of Lancaster House's provisions protecting individual rights.

Dissident activity by ex-ZAPU troops in Matabeleland engendered a harsh government response in early 1983. After more than a year of escalating tension between ZANU and ZAPU, Nkomo fled the country for several months in 1983.

Zimbabwe retains its bright prospects as an economically viable, nonracial, democratic African state, but only if it enjoys enlightened and honest leadership, and if it avoids damage from the increasing level of domestic and international violence emanating from South Africa's political and racial disorder.

8
Accounting for Success

Was the Lancaster House Conference a success? The question is not as improbable as it might seem. The varying perspectives of participants and observers necessitate multiple interpretations. Perhaps it is more analytically helpful to ask whether each of the three major players, allowing for the divisions in two of the delegations, obtained outcomes that met their basic goals.

Britain gained what it had sought unsuccessfully for fifteen years, an honorable way out of the Rhodesian imbroglio. The Patriotic Front obtained, through the electoral process made possible by the conference, what it had been denied on the battlefield, effective political control. Even the Muzorewa delegation, though ultimately vanquished, gained a clear shot at international respectability, the lifting of sanctions, an end to the war, and independence—goals, which though less important than perpetuating themselves in power, were nevertheless major objectives.

The achievements were not without costs. The British were obliged to make commitments they might have preferred to avoid, such as the introduction of U.K. military forces during the transition and promises of aid for the new state. Carrington's reputation was burnished domestically and internationally in the short run. However, some observers believe that the enmity Carrington engendered in his own party's right wing, by foiling their intentions to recognize the Muzorewa government and by his performance at Lancaster House, increased his political vulnerability. In this view, Carrington

paid for his Rhodesian victory with his Falklands resignation,
which Mrs. Thatcher accepted to placate the Tory right. The
Patriotic Front relinquished under pressure many of its fun-
damental tenets during the conference—no special representation for whites, dominant power during the transition, a
supervisory UN role in the ceasefire. As the government of
an independent Zimbabwe, it must operate under a constitution
not entirely of its own choosing. Finally, the Salisbury delegation, and particularly Bishop Muzorewa, paid the heaviest
price, stepping aside for the British governor and then suffering
a humiliating electoral defeat. There were debits, undoubtedly,
but few of the participants, except perhaps the most die-hard
Smith supporters and unconsolable Muzorewa loyalists, would
argue that they outweighed the credits of peace, independence,
return to the world economy, and, for Britain, extrication
from a vexing problem.

Obtaining immediate objectives need not necessarily mean
that a negotiation has established the basis for longer-term
effectiveness or stability. The operation may have been a success,
but everyone knows what happened to the patient. Some
observers argue that Lancaster House ignored the reality of
Zimbabwe's tribal divisions and thus failed to construct insti-
tutional or constitutional mechanisms, such as a federal system,
that would prove to be more equitable and enduring. A more
commonly voiced criticism is that the constitution imposed
upon Zimbabwe has blocked the vast societal restructuring
promised by the Patriotic Front's revolutionary rhetoric, causing an inherently unstable situation in which the will of the
politically mobilized mass finds insufficient gratification. This
view of "the revolution that never was"[1] is more fashionable
outside of Zimbabwe than in it, although it has some currency
among the domestic intelligentsia. A final criticism could be
directed to the conference's negotiating procedure and would
assert that Carrington's highly directive approach brought
about an agreement, particularly the portion relating to the
constitution, that might crumble when the mediator has departed from the scene. A settlement as the product of traditional
bargaining, involving trade-offs and concessions between the

principal protagonists, in this case Muzorewa and the Patriotic Front, would be more likely to endure.[2]

The criticisms are not without merit, but each ignores important facts. The constitution agreed to at Lancaster House created a unitary state with a powerful central government based on the British parliamentary model. It is an inappropriate system for a newly independent heterogeneous society in which the competition for power was, and is, intense. A federal system that would have given greater scope for the channeling of ethnic, personality, and ideological differences into multiple and healthier competitions for power would have been a better solution for Zimbabwe. The only problem with this approach was that none of the African leaders at Lancaster House was interested in such a structure. Nkomo, Mugabe, and Muzorewa were not fighting the battle for black rule in order to lessen governmental authority through federal or other power-sharing systems. More importantly, African self-esteem would not have permitted discussion of constitutional systems that would seemingly have reinforced the white Rhodesian stereotype of festering tribalism mitigated only by white interposition.

The argument which holds that the constitution has unfairly limited the freedom of action of the new Zimbabwean government also ignores certain Lancaster House facts of life. The limitations—the bill of rights, the protection of pensions and private property, etc.—were part of the price the Patriotic Front had to pay to obtain the acquiescence of the Muzorewa delegation, particularly its powerful white contingent. The constitution is far from a perfect document, but it represented a necessary compromise among contending interests. To condemn the bargain that was reached is to rail against the very nature of the negotiation process in which all parties must obtain some satisfaction and no party can be totally pleased.

In relation to the third critique—the possibility of the settlement's crumbling because of insufficient direct negotiations between the protagonists—it must be noted that the Lancaster House agreement has held up relatively well. Moreover, the tensions that exist in today's Zimbabwe between the forces of Nkomo and Mugabe would not likely have been ameliorated by more direct negotiations between the Patriotic Front and the Muzorewa delegation in London. In sum, though criticizable, the conference must be seen as a success in producing outcomes, which while far from ideal, were nevertheless logical, obtainable, and relatively fair.

What then accounted for this success? Carrington has ascribed the Lancaster House agreement to "a combination of luck and circumstance." Several years after the conference he told a reporter:

> The Rhodesians were beginning to worry about their capacity to go on fighting, and about their economy. So they had an incentive to see whether they could settle the thing. And, equally, the leaders of the Patriotic Front were beginning to worry that they weren't going to get possession for a very long time indeed. I don't think they ever doubted they would win in the end, but I think they felt that it was by no means certain they would win in a time span that was acceptable to them. And perhaps just as important from their point of view was the fact that the Front Line states were very bored with the whole exercise. Particularly the Mozambicans, who were suffering a good deal because of it—but so were the Zambians.[3]

Carrington's assessment is only partially correct. Perhaps for reasons of modesty, he focuses only on the objective circumstances surrounding the struggle. He neglects to mention that the changed conditions would not, in themselves, have been sufficient to bring about success. The other element was the negotiating dexterity of Carrington and the British team.

Joshua Nkomo told the press on several occasions that "it is not the Conference that has changed things. It's the circumstances that have changed."[4] To a degree he was correct. The military stalemate, Rhodesia's growing economic squeeze, the desire of both the Front Line states and South Africa to bring the conflict to an end, the advent of an activist British government, and perhaps most importantly, the firm belief shared by the Patriotic Front and Muzorewa that in a fair election cach would emerge victorious over the other, were

all circumstances conducive to a settlement. The essential
ingredients of a successful negotiation, issues of conflict and
a common interest in ending the conflict, were present.[5]
However, without skilled conference management and negotiating expertise all the goodwill in the world would not have
produced a successful outcome. The British edge at Lancaster
House, Carrington's ability to direct the negotiations toward
his desired goals, stemmed from two broad areas of superi-
ority—situational advantage and tactical sophistication. The
first relates to those characteristics of the conference that
largely preceded it or were inherent in the history and character
of the participants. These factors, although givens, were not
seen by the British as static. They were viewed as manipulable.
Manipulation, however, played a greater role in the latter
category of tactical sophistication.

Situational Elements

Clarity of Goals. Despite the suspicions of the Patriotic Front,
there was no hidden British agenda at Lancaster House.
Carrington may have had objectives other than constructing
a peaceful transition to Rhodesian independence—creating a
pro-West Zimbabwe, or choosing the leaders of Zimbabwe's
first government, or limiting Soviet inroads into southern
Africa, etc.—but these were clearly subsidiary and sacrificial
to the overriding goal. A negotiated settlement was in itself
an end, not a means to something else. Carrington, unlike
Henry Kissinger in the Middle East, did not operate within a
global or regional strategic design into which a Rhodesian
settlement had to fit.

Absence of Superpower Involvement. Rhodesia, unlike Angola,
had not developed into a cockpit of U.S.-USSR rivalry. Potential
superpower conflict had been mitigated by the Carter administration's conscious decision to cast the Rhodesian problem
in an African, rather than East-West, context.[6] Although both
the Soviet Union and China served as arms suppliers to the
guerrilla movements, the competition between these two powers, and, for reasons quite unrelated to the Sino-Soviet dispute,

the tension between ZANU and ZAPU effectively prevented either of the Communist colossi from obtaining any significant influence over their respective clients in the Patriotic Front. The absence of a larger geopolitical construct was reinforced by the British decision at the outset to maintain its centrality in the negotiating process, relegating the United States to the role of interested observer and supporting actor. Nor was either element of the Patriotic Front interested in converting the conflict and its resolution into a sideshow of East-West rivalry. Genuine sentiments of African nationalism, Front Line and Commonwealth involvement, the fear of greater South African intervention, and Mugabe's deep distrust of the Soviets (who had continually pressed him to accept Nkomo's suzerainty) all conspired against further internationalization of the conflict. Neither the Soviet Union nor China had a role to play in the London deliberations.[7]

The British Diplomatic Heritage. Some observers note elements of Britain's long-evolving political culture, which, they argue, make that country's diplomats particularly adept at negotiation. Sir William Hayter wrote that as Britain is ruled by committees, the most important being the cabinet, British diplomats are well schooled in the manipulation of conferences.[8] Nicolson ascribes his countrymen's negotiating success to

> moments of enlightenment when they recognize that the art of negotiation is essentially a mercantile art, and the success of British diplomacy is to be explained by the fact that it is founded on the sound business principles of moderation, fair-dealing, reasonableness, credit, compromise, and a distrust of all surprises of sensational extremes.[9]

It is certainly debatable whether British negotiators have necessarily been any more successful than those of other nations, or, if so, whether any one cultural factor—be it sharp committee practice or a shopkeeper's mentality—can account for the success. In any event, there are few pure personality types—every shopkeeper has a little warrior in him.[10] What is indisputable is that Britain has been at the diplomatic game

for centuries. The residue of this experience forms a negotiating tradition that is likely to be more coherent and effective
than the negotiating styles of others with less international
experience. And even if it is not, its very appearance—
Whitehall's weathered facade, Lancaster House's ornate conference room—can convey an impression of success, wisdom,
and experience, which may, or may not, be accurate.

The Conservative Party Tradition. The Tories have consistently
given their leaders more latitude in the conduct of foreign
affairs than has Labour. Moreover, Conservative political figures have shown themselves to be relatively hardheaded, rather
than idealistic, to use Nicolson's term, about Britain's overseas
interests. They have sought to enhance these interests in consort
with, rather than in opposition to, Whitehall's professional
diplomats. Although these qualities do not guarantee wise
policies, they do provide for a unity of purpose that makes
negotiation easier. Carrington and his staff shared a mutual
respect. Mrs. Thatcher's unchallenged Tory leadership and
demonstrable trust in her foreign secretary allowed the British
to present a united front at Lancaster House—an obvious
advantage over the fractionated Patriotic Front and Muzorewa
delegations.

Carrington as Symbol. As a Tory, Carrington had other
advantages. The Muzorewa delegation realized that it was
likely to obtain a better deal from a Conservative government
than from a Labour one. This view was supported by the
historical record—the almost successful Smith-Home agreement, right-wing Tory support for Smith and then for Muzorewa, and Mrs. Thatcher's expression of Conservative an-
tipathy toward "Marxist terrorists." But there were other
elements of the Tory tradition that engendered a special appeal
for both white and black Rhodesians. Most Rhodesian whites,
many of whom had emigrated from the U.K. shortly after
World War II, felt a general disdain for "socialist," empire-
less, no-longer-a-world-power Britain. To them Carrington, a
hereditary peer in the service of an aggressive, almost imperial,
Mrs. Thatcher, was a comforting figure whom they could
respect. The African response to the Tory tradition as em-

bodied in Carrington was considerably more complex, reflecting the inherent ambivalences of colonialism. At base, however, Carrington's social and economic position conformed to the African stereotype of those who continue to wield power behind Britain's facade of modern egalitarianism. Thus, he was a worthy negotiator. Moreover, Africans on both delegations may have viewed Carrington and his staff of relatively conservative (some might say stuffy) Foreign and Commonwealth Office types with less suspicion than they reserved for such Labour figures as David Owen and Geneva Conference chairman Ivor Richards, whose attempts at bonhomie, based on the shared principles of anticolonialism and socialism, often somehow fell flat. "Perhaps," as one Patriotic Front delegate commented near the conference's close, "the Tories are the natural party of decolonization."[11]

Carrington's Qualities. The foreign secretary possessed many of the qualities of a good negotiator: self-assurance, tact, dignity, humor, ability to demonstrate anger, intelligence, and a personal magnetism variously described as lordly charm or aristocratic bearing. He had significant, if not totally satisfactory, experience as a party official, having served as the Tory chairman in the mid-1950s, and though "he had never fought an election or heard the rattle of live ammunition in the Commons,"[12] he had a keen sense of what was politically acceptable. As a former minister of defense and high commissioner to Australia he was internationally schooled. As the principal negotiator in 1971 facing Dom Mintoff on the Maltese base issue, he had earned his negotiating merit badge dueling with the craftiest bargainer this side of a souk. In addition to his qualities and training, Carrington's most important characteristics were his considerable capacity for hard work, his mastery of detail, and his ubiquity. His continuous presence provided continuity in leadership and a personal manifestation of his government's commitment to a negotiated settlement.

The Setting. Lancaster House had been the scene of prior British constitutional conferences that established the terms for granting independence to several of its colonies. Its selection was purposeful and designed to convey the clear message that the conference would not be just another in a series of talks, but rather the final act in the Rhodesian drama and the first scene of Zimbabwe's history. In this regard Lancaster House was an example of innovative negotiations described by Ikle, which are designed to create a new relationship or new undertakings among parties.[13]

The risk of exclusion from the to-be-created Zimbabwean political dispensation weighed heavily on the Patriotic Front and was a principal factor in confounding those who had predicted their walkout from the conference. Apart from the symbolism, the London site allowed Carrington full and easy access to the panoply of tools at his disposal—the political support of the prime minister, the parliamentary forum, the FCO bureaucracy, and informed officials of the Commonwealth and other governments. The British had the home field advantage.

The Press. Flowing from the site selection was Carrington's ability to use the press effectively. Journalists naturally gravitated to the FCO for authoritative information on the conference's proceedings, which was provided by Nic Fenn, acerbic and articulate British press officer. Fenn had the best of both worlds. He would brief the press on background as the conference's spokesman, giving a British slant to developments, and then tell the journalists, "Now my British colleague is at your disposal to answer questions."[14] Fenn would then continue, wearing his Carrington's spokesman hat. Ready access to the press also allowed the British to leak information in pursuit of their own tactical advantage, as evidenced by repeated FCO-stimulated articles about the imminence of the second-class solution or dissension within the Patriotic Front's ranks.

Intelligence. The *Sunday Times*'s titillating revelations of taping and bugging of the Patriotic Front's hotel rooms and telephones have been overplayed by some observers.[15] The delegates' telephone talk was often guarded, and thus of limited value. But there is no question that the information available to Carrington was first-rate. Much of it was obtained overtly and covertly from established sources within the other delegations.

The information gleaned from all sources was funneled to Carrington each morning.

In experiments with inexperienced negotiators one researcher found that giving them information about their opponents tended to create "a morally neutral climate" that undermined traditional values of cooperativeness.[16] Carrington and his staff had been reading other gentlemen's mail for some time, and it is doubtful that snooping robbed them of their innocence. However, the information gathered may have increased British self-confidence and fostered a tougher negotiating attitude. At the least, it placed Carrington in a better position to decide when additional pressure or concessionary moves were in order.

Perceived Partiality. Before, during, and after the conference, Britain was perceived by the other parties as biased in favor of the Muzorewa regime. Traditionally, impartiality has been considered the sine qua non of effective mediation.[17] In reality, however, a mediator who is thought to be biased may, in some circumstances, bring to a negotiation qualities appreciated by all the parties, including those against whom a bias is perceived.[18] Moreover, the bias itself becomes an instrumental factor in negotiating dynamics. The "favored" party seeks to maintain its good relations with the mediator and is, thus, more prone to making concessions. Clearly, the Salisbury delegation saw amicable relations with Carrington as a means of maximizing its conference outcomes: initially the lifting of sanctions and recognition; later continued U.K. beneficence during the transition period. This attitude dovetailed with the British view, which maintained that the trauma of relinquishing power must be compensated for in several ways, not the least of which is enhanced personal relationships with the mediator.

Certainly, the Patriotic Front would have preferred a less partisan (in their view) third party, or, even more desirable, a mediator biased in their favor. For them, the greatest hazard posed by a biased mediator was the bargaining leverage that the threat of coalition of the mediator and the favored party, that is, the second-class option, provided Carrington. But even for the "unfavored" party a mediator who is perceived as partisan is somewhat attractive, if he can build upon the rapport with the "favored" party to obtain concessions that would otherwise be unattainable. It is arguable which delegation conceded more at Lancaster House, but ultimately the most significant concession was the bishop's stepping aside for a British governor and new elections. Nkomo and Mugabe realized that acting alone they could not have obtained that concession short of a military victory, which still appeared distant.

Carrington's Tactics

The circumstances surrounding the struggle and Lancaster House's situational elements were conducive to a satisfactory outcome. They made "the fruit ripe for picking," as Nkomo said. But they were only the necessary, and not sufficient, conditions. Carrington's tactics provided the other element of success.

The Step-by-Step Approach. The keystone of the British conference tactics, the step-by-step approach, generated momentum as intended, kept the parties engaged, and conveyed the impression of conference progress necessary for Carrington to maintain the support of Mrs. Thatcher and interested onlookers such as the Front Line states and the U.S. government. Momentum itself became a conference subgoal, as illustrated by the Patriotic Front's agreement to bifurcate the final discussions on the ceasefire so as to demonstrate progress at a time when a second-class solution or a Salisbury walkout was being threatened. The tactic also enabled the British to maintain issue control by placing those topics more amenable to agreement, such as the constitution, closer to the top of the agenda. "Fractionating the conflict" seems almost invariably the best tactic for a negotiator seeking to bring about change.[19] Salami tactics, in which a problem is taken one slice at a time, presents to a potentially obstructionist opponent less of a target at which to aim.

There was nothing particularly subtle about the British approach. One conference observer termed Carrington's approach the "gag and swallow" method: "take this piece, choke if you must, but accept it, so we can move ahead." All conference participants knew what was going on. Those least inclined to a settlement were the most reluctant to swallow. Smith justified his vote rejecting the constitutional proposals be telling the press that "I don't accept something that is dependent on something else happening in the future. I want to see the whole deal clear."[20] His views were echoed a few days later by a recalcitrant Patriotic Front spokesman who said, "We don't want to do anything under duress. It is not rational to expect us to give this reply before we have the total picture of the agreement."[21] Nevertheless, they repeatedly did so, weakening their own bargaining position for the next phase.

Manipulation of the Supporting Players. The preconference determination to reaffirm Britain's central role in the negotiations and to remove nonparticipants from the center to the periphery in no way lessened the supporting actors' interest in the negotiating process. Thus, although the Americans were no longer the equal partners that they had been for the previous three years, they were still able to offer assistance at appropriate junctures, e.g., the aid pledge and the lifting of sanctions when requested. Similarly, the Front Line states and the Commonwealth were called upon for support and mediating effort when necessary.

A principal function of the supporting players was to reinforce Britain's credibility as an honest negotiating partner. "One of our most important contributions throughout the Lancaster House Conference," writes Cyrus Vance, "was to vouch for British sincerity and impartiality with the suspicious Africans."[22] The Front Line states and the Commonwealth adopted a similar posture with the Patriotic Front, while the South African government played an analogous role in relation to the Salisbury delegation.

The ironies of this approach were multiple. First, some of the supporting actors, notably the South Africans, were hopeful that Britain's negotiating sincerity would be skewed in such a way as to favor the Muzorewa team. Second, many of those proclaiming the rectitude of the British harbored secret doubts, which they were willing to subordinate in pursuit of the shared goal of a negotiated settlement. Finally, the image of British impartiality was not always consistent with Carrington's tactics, which derived negotiating strength from the carefully cultivated impression of the Conservative government's bias in favor of Muzorewa.

Another important role of the supporting actors was to serve as channels of communication and trial balloon floaters for the three delegations, allowing new ideas to circulate without any delegation's formally accepting paternity. The British were the most skillful in using intermediaries, particularly Ramphal and Front Line representatives. Moreover, Carrington was not averse to utilizing the peripheral actors for his intragovernmental negotiations. Though not documentable, it is more than a fair bet that some of the arguments advanced to Mrs. Thatcher directly, or through Carrington, from Ramphal, the Carter administration, or others were instigated by the foreign secretary himself or his staff.

Of course, not all of the interventions of the supporting actors were welcomed by the British. But on the whole, British centrality helped to create an encircling wall around the negotiators. Like a Chinese finger puzzle that tightens as the victim tries to extricate himself, the wall closed in on the principal actors as they pushed on it for support. The net effect was the creation of a corps of kibitzers dedicated to seeking a successful resolution, though, of course, they may have differed among themselves in their interpretation of success.

Insistence on Carrington's Centrality. Carrington arrogated for himself the central role. Prior to the conference, the British established that there could be no appeal from his decisions, no way of going around him to seek reversal. Significantly, throughout the three-and-a-half months of the conference, Mrs. Thatcher did not meet once with Nkomo, Mugabe, or Muzorewa. Their positions were undoubtedly touted to her by Ramphal, Kaunda, Botha, and others, but she continually made it clear that her foreign secretary had her complete confidence and total backing. Moreover, the structure of the

conference, in which the Patriotic Front and Salisbury negotiated with Carrington rather than with each other, reinforced his centrality. It is unknown whether he would have
objected to direct Patriotic Front–Salisbury talks, if Muzorewa
had approved of them, but clearly the process that did evolve
best suited British purposes and probably those of the other
delegations as well.[23] In effect, Carrington and his team engaged
in shuttle diplomacy of a sort. Though the parties sat in the
same room for plenary meetings, no bargaining took place in
those position-stating sessions, leaving it to Carrington to carry
the negotiating load. In this role, he was able to obtain from
each delegation concessions that they would not have granted
to the other. Carrington's crucial role was also reinforced by
his insistence on basing the negotiations on his texts, forcing
the opposing parties to focus on views other than their own.[24]

Huff and Puff and Fudge.[25] Carrington was able to create a
sense of conference pressure (utilized principally on the Patriotic Front) by setting deadlines for his proposals. It says
much about his air of self-assurance that he was able to ignore
repeatedly his own "ultimatums," as the press enjoyed calling
them, without ever giving the impression of having backed
down. In point of fact, the British negotiating style coupled
a harsh public posture with a far more nuanced and sensitive
private relationship with the Patriotic Front. Doors were never
slammed shut. Much of the tough negotiator facade was pure
theater, designed to have its effect on Nkomo and Mugabe,
but also intended to mitigate right-wing Tory, South African,
and Salisbury charges that Carrington was treating the Patriotic
Front too gently.

Tacit Bargaining. The British engineered a series of extra-
conference moves designed to force the pace of the Patriotic
Front's acquiescence to Carrington's successive proposals. The
discussion of transitional arrangements was speeded to a close
by the introduction of enabling legislation that might permit
a second-class solution. The December 3 Order in Council,
giving Mrs. Thatcher the specific authority to select a governor,
spurred the Patriotic Front to accept most of the ceasefire

plan. The sending of Soames to Rhodesia prior to final agreement was the most audacious piece of extra-conference pressuring, but one the British felt necessary to bring the proceedings to their end. Other elements of Carrington's tacit bargaining were the effective utilization of press conferences, sanctioned leaks, and parliamentary debates to make negotiating points that might have seemed out of place in the confines of Lancaster House.

British ability in this area clearly overshadowed that of the two other delegations. The Salisbury team's effort to affect the negotiations through close association with right-wing Tories was neutralized by Carrington, as evidenced by his performance at the Blackpool conference. Rhodesian military attacks into Mozambique and Zambia never conveyed a clear conference-related message. And the Salisbury threats to leave London were handled relatively easily by Carrington, who turned them to his advantage in increasing pressure on Nkomo and Mugabe. The Patriotic Front sought to make use of the press and engaged in symbolic acts such as boycotting the opening night reception and flying to Dar es Salaam to obtain a vote of Front Line confidence. But their efforts were neither as sustained nor as effective as Carrington's.

Compellance. Carrington's principal bargaining leverage over the Patriotic Front was supplied by a skillful and conscious policy based on the manipulation of compellant threats. As defined by Schelling,[26] compellance involves making an irrevocable commitment to action that can cease or become harmless only if the opponent responds as specifically desired. As with any coercive threat, compellance denotes a recognition that the punishment will damage the threatener as well as the threatenee. Further, the threat itself requires an accompanying assurance that it will not be acted upon if the requested responding action is undertaken. The second-class solution constituted the compelling threat against the Patriotic Front.

Fisher and Ury note that a negotiator measurably strengthens his position by developing an understanding of his BATNA— best alternative to a negotiated agreement.

> People think of negotiating power as being determined by resources like wealth, political connections, physical strength, friends, and military might. In fact, the relative negotiating power of two parties depends primarily upon how attractive to each is the option of not reaching agreement.[27]

They note that it is often desirable for a negotiator to let his opponent know what his BATNA is, especially if it is a plausible alternative that would provide the negotiator with benefits, albeit less than those to be obtained from a successful conclusion of the negotiation. At Lancaster House, Carrington not only let the Patriotic Front, and everyone else, know his BATNA, the second-class solution, but he was able to use it as an instrument of considerable leverage. The press repeatedly referred to Carrington's tactics as "brinkmanship." It may, however, not be an accurate description, for brinkmanship denotes a willingness to manipulate the risk of mutual disaster for negotiating leverage. It is unclear whether Carrington viewed a second-class solution as a disaster for Britain. Certainly, he recognized, and the Patriotic Front understood that he recognized, that such a solution would be messy and potentially dangerous to Britain's prestige. But Carrington was able to make the threat a credible one by his extra-conference maneuvers, which demonstrated commitment, a British binding of themselves to the threatened action. The ultimate dispatch of Soames to Rhodesia was a classic bridge-burning tactic which signaled that there could be no turning back from implementation of the compelling threat unless the Patriotic Front accepted the Lancaster House agreement.

In pursuing the tactic of compellance the foreign secretary was aware that it does not always help to be thought of as totally in control of events. He reinforced with the Patriotic Front the impression that he was under pressure that could force him into the second-class solution. In this regard, continued right-wing Tory sniping played into Carrington's hands. His leverage was also increased by keeping the Patriotic Front from direct contact with Mrs. Thatcher and by telling Nkomo and Mugabe, not inaccurately, that his boss was losing patience with them and might choose to act less rationally than he.

An interesting missing element was that the Patriotic Front never chose to threaten hostilities against Britain, even though resumption of British authority under a second-class solution might have placed them in rebellion against the Crown. There were, of course, occasional rhetorical flourishes such as ZANU spokesman Zvobgo's promise of an all-out war against a British governor should the conference fail. But this line was never seriously pursued. Could Carrington have wielded the club of a second-class solution so sanguinely and with as much Tory support if the Patriotic Front had done more saber rattling? Certainly not. Then why did not Nkomo and Mugabe adopt a more threatening posture? The answer is complex but reveals many of the same factors that inhibited them from seeking Soviet or Cuban combat support—Front Line and Commonwealth desire to avoid further internationalization of the conflict, prior acknowledgment of Britain's legitimate sovereignty over Rhodesia, and a commonsensical recognition that a more threatening stance would destroy whatever slim likelihood may have existed of the conference's reaching an acceptable agreement.

Promises. Schelling notes that the exact definition of a promise, in distinction to a threat, is not obvious. In fact, a threat depends on its specificity for its effect. A promise's effect often increases with its vagueness. Threats were not a significant element in the British approach to the Muzorewa delegation, but promises or the perception of them were. Immediately following the elections, General Walls sent a cable to Mrs. Thatcher asking her to invalidate the results on the basis of widespread Patriotic Front intimidation of voters. Behind that demand, which was ignored and did not become public for many months, was a profound sense of betrayal, palpable in Rhodesia among many whites and Muzorewa supporters. No hard evidence has ever been produced to indicate that the British promised a different electoral outcome. It is doubtful that Carrington or others would have placed themselves or their government in the compromising position such a pledge would entail. Certainly, the British promised the Muzorewa delegation that their interests would be protected during the transition—and this was no small incentive for their acceptance of Carrington's proposals. But did the British promise "to see the Bishop right" mean different things to the two parties? Probably. Did the British scrupulously seek to correct any misimpression their words may have conveyed to the bishop's eager ears? Probably not. Muzorewa, Walls, and others would have done well to read what another Briton had written about people hearing what they want to hear:

> Her speech is nothing,
> Yet the unshaped use of it doth move
> The hearers to collection; they aim at it,
> And botch the words to fit to their own thoughts,
> Which, as her winks, and nods, and gestures yield them,
> Indeed would make one think there might be thought . . .[28]

9
Third Encounters of the Close Kind: Dominant Third-Party Mediation

Students of conflict resolution might find it helpful to combine the study of the Lancaster House Conference with interpretations of other negotiations, successful or otherwise, in order to draw conclusions or develop hypotheses that go beyond this one set of experiences. The following thoughts are offered not as a coherent theory, but rather as ideas that might guide further research.

It is difficult to describe the British role at Lancaster House. Was Carrington a mediator? By all means. He took the opposing views of the Patriotic Front and the Salisbury delegations and by dint of skill, circumstance, and tactic found common ground on which both could agree. But wasn't he really more of a negotiator than a mediator? By all means. The two other delegations did not bargain in any real sense with each other, but only with Carrington. He devised proposals, engaged in tacit bargaining, issued threats, murmured promises—in short, an able and effective negotiator. On the other hand, couldn't he be seen as an arbitrator, listening to both sides, then issuing his own solution, which the others were obliged to accept? By all means . . . and so on.

The categorical difficulty lies not entirely in the multiplicity of Carrington's roles. It is also prompted by the blurring of distinctions that were once thought to neatly exist. The pigeons have come home to roost and the pigeonholes are in a frightful state. "Certain subjects seem quite clear as long as we leave them alone," comments Ikle.[1] Truer words were never written.

Not all conflicts may be solvable. Indeed some need not be dealt with at all. Stalemate prevails or fate intervenes: men die; governments fall; the irreconcilable becomes the irrelevant. However, there are conflicts that must be confronted. They can be resolved by dictation, negotiation, or mediation. But once we start poking around we find that these processes are not easily separated into discrete entities.

Dictation occurs when one party to a dispute imposes its will on the other. War is a classic form of attempted dictation, but contemporary theorists tell us that modern warfare is itself a form of negotiation, marked by tacit bargaining designed to limit or terminate hostilities. Similarly, the distinction between negotiation and mediation is often blurred. In multilateral fora negotiators will become mediators to resolve differences among other negotiators. On occasion, ego or a sense of morality or of justice might prompt a negotiator to mediate between his opponent and his own government or sponsor. Indeed, one school of thought asserts that the more negotiators act like mediators—objectively sifting the interests and principles at issue—the more successful they are likely to be.[2] Finally, mediation blurs not only into negotiation, but on occasion tends toward dictation.

The ideal mediator was once thought to be an impartial third party, with no particular stake in the outcome, able to elucidate the conflictive issues, promote cooperation among the parties, and generate compromise agreements. Most mediators still perform these functions. In international relations, however, fastidious solomonic eunuchs are in short supply. Many mediators represent entities that are not disinterested and have the capability of influencing participants by the use of threats and promises. If this "mediator with muscle" had the requisite force, will, and freedom of action to impose a settlement, he might become a dictator.[3] But in a world of autonomous states and rampant nationalism, dictation is rarely possible.

It is perhaps best to see mediators as operating on several continua at once. One extends from nonaggressive information gathering and message passing at one end to highly intrusive manipulation of issues and actors at the other. Another continuum ranges from relative disinterestedness to consuming concern in the outcome. And a third runs from relative powerlessness, in which the mediator has little influence or leverage, to extreme power, in which the mediator's position is stronger than that of any of the more directly affected principals. Mediators may move back and forth along each continuum during the course of a negotiation. When a third party operates at the far end of all three continua, that is, when he is highly manipulative, is very much concerned about the outcome, and has much leverage over the other parties, the necessary, but not sufficient, conditions exist for a special kind of mediation.

This type, perhaps not common, but on the other hand not unique to Lancaster House, lies between mediation with muscle and dictation. We shall call it dominant third-party mediation. It can be described with a homely example. Two children argue about which film the family will see on Saturday. One asserts the absolute indispensability of seeing E.T. for the seventh time. The other argues the case for Rocky XV. Unable to agree, they call in an older brother, but his advice is not accepted. Visiting Uncle Henry tries to end the dispute. He promises differential rewards at Christmastime or threatens, in exasperation, no birthday presents. But neither Christmas nor birthdays are more important just now to the children than winning the movie battle. Enter the father. Dictation is considered. "Shut up," he could explain, "no movie at all this weekend." But that would not serve his purposes: he is not prepared to listen to the family complain; there will be hell to pay with his wife. So far negotiation, mediation, and Uncle Henry's mediation with muscle have failed. Dictation is not practicable. It is at this juncture that the father becomes a dominant third-party mediator. "I've got the keys to the car and the money to buy the tickets. Unless you come to an agreement that satisfies me (I've got to sit through the film as well), there will be no movie, that is, no settlement. Now here's my idea of how we can compromise. . . ." All the players accept that the father has the ultimate authority to implement an agreement, and, therefore, has the right to assume the principal role in bringing about a settlement. But his power is not absolute. He needs the kids' cooperation. They may not accept all of his ideas. They may come up with some of their own. They can negotiate, but they recognize that his imprimatur is indispensable.[4]

Carrington was the father at Lancaster House. The very term is almost embarrassingly out of place in the modern world. The conference was remarkable in recent history in that the dominant third-party mediation was so overt. That, however, was attributable to the nature of the dispute and the conference that ended it. It was a decolonizing venture— perhaps the last to be seen—in which the British used the skills of manipulation honed over centuries of imperial rule. Few other examples of visible dominant third-party mediation can be found in a world that frowns on displays of authority by one sovereign state over another. It is, therefore, difficult to point to specific negotiations that parallel Lancaster House in this regard. Hypothetically, however, one might ask what the role of the Soviet Union would be in negotiations designed to end a serious trade dispute between Bulgaria and Hungary that might affect Russian interests. Or what role would France play in settling a similar dispute between members of the West African French franc zone?

Outside the sphere of international relations, dominant third-party mediation might be more visible and thus more easily studied. For instance, what is the role of national governments when workers strike against the supposedly autonomous management of a state-owned corporation? Or how do we describe the intervention of a "jawboning" government in major negotiations between producers and consumers designed to set prices for principal products—steel, for example? The list could be extended with instances of potential dominant third-party mediation in commercial, industrial, judicial, and familial areas as well as in the international arena.

The role of the nondisinterested powerful mediator has only recently attracted scholarly attention and as yet there is no comprehensive theory that adequately describes the phenomenon. One will not be presented here. However, on the basis of the Lancaster House experience certain abiding characteristics of dominant third-party mediation may be postulated. The dividing line between this form of mediation and its closest cousin, mediation with muscle, is not always clear. In fact, the principal actors in both conduct themselves in much the same fashion. The tactics employed in Kissinger's muscled mediation in the Middle East did not differ greatly, in most regards, from those employed in Carrington's dominant mediation at Lancaster House. Both were at the far ends of the three continua. The distinction rests not on their functions (what they did) but rather on their roles (who they were).

The key to understanding dominant third-party mediation lies in revamping somewhat accepted definitions. Jeffrey Rubin, in his excellent introduction to a fine collection of essays on Kissinger's involvement in the Middle East, writes:

> Indeed by virtue of its very semantic construction, the term "third party" implies the existence, the reality, of at least two others about which several things appear evident: the principals were there first (namely, first and second). Had they not been there first, there would have been no third; the third party is thus spawned by the relationship between the other two. Additionally, because the principals were there first, their exchange takes precedence over any relationship that the third party may have with either of them. The role and involvement of a third party are thus typically peripheral to the primary relationship.

He immediately acknowledges that the third party can assume transcendent importance and alter greatly the relationship between the first two, perhaps becoming the only cement holding it together, but "even here, the third party must be seen as an outgrowth of the relationship between the first two."

Rubin's characterization of the "third party" as a product of the first two does not adequately characterize dominant third-party mediation. In fact, the reverse is true. At Lancaster House, the participants had no relevant existence apart from their relationship to the British. The third party created the first and second for the purposes of negotiation. Egypt and Israel theoretically could have negotiated a settlement without outside involvement. Their practical inability to do so gave the mediator more opportunity to flex his muscles, but the mediator was not indispensable. In the Rhodesian situation, on the other hand, the centrality of Britain was clear and the theoretical as well as practical impossibility of obtaining a negotiated settlement without British involvement was recognized by all.

It is toward this distinction that further study should be directed. What makes dominant third-party mediation possible is the recognizable authority of the third party. The necessary condition precedent is an existing relationship between the disputants on the one hand and the mediator on the other, which accords the third party a readily accepted status as an indispensable intervenor.

This formal role, whether de facto or de jure, is critical. At Lancaster House all parties and observers accepted as given the inherent right and authority of Britain as the only entity capable of certifying a settlement through the granting of independence. This unchallenged position allowed, but did not make inevitable, the exercise of British power that brought about the settlement. The same recognition of Britain's authority had existed since Smith declared his unilateral independence. What distinguished Lancaster House from previous negotiating efforts was the willingness of Mrs. Thatcher and Carrington to devote themselves fully to a settlement, in other words, to extend the British government to the limits of the three continua. But even this commitment would have yielded nothing without Carrington's negotiating skill, which, in turn, would have been ineffective if the conditions external to the conference had not been present.

In sum, then, scholars and observers might search for additional examples of dominant third-party medition at that conjuncture where circumstances conducive to a settlement interact with a third party with recognized authority to intervene in the dispute and with sufficient aggressiveness, interest, power, and skill to do so effectively. How much power is necessary or likely to be utilized is variable and, at least for now, indefinable, but it probably lies somewhere between the amount necessary for mediation with muscle and the quantity sufficient to permit dictation.

The topic warrants further exploration. As one U.S. diplomat who played an important role at an earlier stage in the search for a Rhodesian settlement has recently written:

> Conflict is the one predictably certain element in future relations between nations, particularly in the third world. As long as the United States continues to maintain its position of enormous world influence in the eyes of others, as it certainly will for the foreseeable future, it is condemned to act in the role of mediator almost steadily. It simply has no choice but to offer peace plans in situations like the Middle East and southern Africa and follow them up with energetic, good faith negotiations. No major dispute in Latin America, Africa, the Middle East, Europe or the Pacific is likely to arise without early appeals to the U.S. to exert its influence to prevent or resolve it.[5]

Mediation, in all its variations, will be a recurring feature of the diplomacy of the United States as well as other countries. It might help to know a bit more about it.

Notes

Chapter 2: Setting the Stage

1. For an excellent account of the impact of sanctions on Rhodesia, see Robin Renwick, *Economic Sanctions* (Cambridge: Harvard University, Center for International Affairs, 1981).

2. Alex Callinicos, *Southern Africa After Zimbabwe* (London: Pluto Press, 1981).

3. Though called *Zimbabwe-Rhodesia* by the Muzorewa government and *Zimbabwe* by the Patriotic Front, the country will be referred to in this book as Rhodesia for the period of the Lancaster House talks. The "legal" name actually remained the Colony of Southern Rhodesia.

4. The first five of the six principles had been developed by Prime Minister Harold Wilson in 1964, before UDI. They follow:

1. The principle and intention of unimpeded progress to majority rule, already enshrined in the 1961 constitution, would have to be maintained and guaranteed.
2. There would also have to be guarantees against retrogressive amendment of the constitution.
3. There would have to be immediate improvement in the political status of the African population.
4. There would have to be progress towards ending racial discrimination.
5. The British government would need to be satisfied that any basic proposal for independence was acceptable to the people of Rhodesia as a whole.

A sixth principle was added by Wilson in 1966.

6. The need to ensure that, regardless of race, there is no oppression of the majority by the minority or of the minority by the majority.

By 1979 many of the principles had been overtaken by events, but the concept of satisfying the six principles had become an inviolable part of succeeding British governments' Rhodesia policy. The portion of the Conservative party manifesto dealing with Rhodesia is quoted in Miles Hudson, *Triumph or Tragedy* (London: Hamish-Hamilton, 1981), p. 148.

5. Martin Meredith, *The Past Is Another Country* (London: Pan Books, 1980), p. 364.

6. Hudson, pp. 156–157.

7. John Newhouse, "Profiles (Lord Carrington)," *New Yorker*, February 14, 1983, p. 55.

8. Ibid., p. 71.

9. Hudson, pp. 156–157.

10. An account of the May 21 meeting appears in Cyrus Vance, *Hard Choices* (New York: Simon and Schuster, 1983), pp. 295–296.

11. Unlike Vance, who devotes two chapters of his memoirs to southern Africa, Brzezinski had little interest in the area. The following entry in his July 1977 diary (that is, two years before Vance's meeting with Carrington) reveals a more accepting attitude toward Muzorewa than that prevalent in the State Department.

> I am beginning to lean to the notion that we ought to let the so-called internal solution surface and let the moderate Africans take over from Smith because it is only to them that Smith can yield, and then let the internal solution based on the moderate leadership collapse as the more assertive Africans storm in from the outside. If we can maintain a policy of benevolent neutrality, we can accelerate the process and maintain a positive relationhsip with the Front Line African states which ought to be the central focus of our efforts.

Zbigniew Brzezinski, *Power and Principle* (New York: Farrar, Straus & Giroux, 1983), pp. 140–141.

12. *Newsweek*, August 13, 1979, p. 38.

13. In the schema devised by Michael Handel ("Surprise and Change in International Politics," *Journal of International Security* 4 [4], 1980), Mrs. Thatcher's actions at Lusaka can be seen as constituting a "minor surprise," defined as an unexpected move that is intended to change the trend in relations between two or more states, although

it does not have a radical impact on the balance of power in the international system. Such moves are easiest for leaders who enjoy complete control over their governments (Hitler, Stalin) but not impossible for leaders with authoritarian personalities in democratic societies (e.g., Nixon, de Gaulle, and, in this instance, Mrs. Thatcher).

14. *Journal of Southern African Affairs* 4(4), October 1979, p. 402. Hereinafter referred to as *JSAA.*

Chapter 3: The Delegations

1. For an interesting review of Tory-FCO relations see Gillian Peele, "The Changed Character of British Foreign and Security Policy," *Journal of International Security* 4(4), 1980.

2. Newhouse, p. 51.

3. See "Up for Auction: Malta Bargains with Great Britain, 1971," by W. Howard Wriggins, in I. William Zartman, *The 50% Solution* (New York: Anchor Books, 1976), pp. 208–234, for an account of those frustrating negotiations from which Carrington drew some of the lessons implemented at Lancaster House.

4. Robin Renwick, "The Rhodesia Settlement," unpublished, Harvard University, Center for International Affairs, 1981.

5. Lord Christopher Soames, "Rhodesia to Zimbabwe," *International Affairs* 56(3), Summer 1980.

6. This quote appears in Diana Mitchell, *African Nationalist Leaders in Zimbabwe: Who's Who 1980* (Salisbury: Diana Mitchell, 1980), p. 99. The volume is an excellent compilation of biographical detail about Zimbabwe's leaders, and contains information that is not readily available elsewhere.

7. See Denis Hills, *The Last Days of White Rhodesia* (London: Chatto and Windus, 1981), pp. 140–141.

8. A point fairly made by Xan Smiley, in "Zimbabwe, Southern Africa and the Rise of Robert Mugabe," *Foreign Affairs* 58(5), Summer 1980, p. 1063.

9. The Front Line states are Zambia, Tanzania, Mozambique, Angola, and Botswana. Nigeria, by virtue of its size, considered itself, and came to be considered, a Front Line state as well. Angola, beset with its own problems, did not take an active role in the Rhodesian negotiations.

10. The sum is quoted in Robert Jaster, *A Regional Security Role for Africa's Front-Line States: Experience and Prospects,* Adelphi Paper 180, International Institute for Strategic Studies, London, 1983, p.

9. Jaster's monograph is a helpful study of the Front Line–Patriotic Front relationship.

11. Shiva Naipaul, *North of South* (London: Penguin Books, 1980), p. 320.

12. See, for instance, *Africa Confidential* 20(18), September 5, 1979, which expressed doubt about Mozambique's interest in pursuing a settlement.

13. For an interesting account of Mozambican attitudes toward ZANU, see David Martin and Phyllis Johnson, *The Struggle for Zimbabwe* (London: Faber and Faber, 1981), pp. 316–318.

14. The *Economist*, September 15, 1979, pp. 15–16.

15. See, for instance, Robert Rotberg, *Suffer the Future* (Cambridge: Harvard University Press, 1980), pp. 264–265.

16. See Martin and Johnson, *The Struggle for Zimbabwe*, for an account of the military's decision to support Mugabe, pp. 191–214.

Chapter 4: Overture and the Constitution

1. The text of the invitation appears in *JSAA* 4(4), p. 404.

2. The text of the eleven principles appears in *African Index*, October 1–15, 1979.

3. This account of the Havana meeting is taken from Robert Jaster's excellent study of the Front Line states.

4. *Financial Times*, September 11, 1979.

5. *Report of the Constitutional Conference* (London: Her Majesty's Stationery Office [hereafter referred to as *Cmnd.* 7802]), pp. 3–8.

6. *Cmnd.* 7802, pp. 9–12.

7. *Cmnd.* 7802, pp. 12–16.

8. *Financial Times*, September 13, and 14, 1979, contains good summaries of the British and Patriotic Front proposals.

9. After Mugabe assumed power, Kamusikiri returned to the United States, where he had lived for sixteen years. Mundawarara was elected to parliament on the bishop's ticket in the independence elections.

10. *Financial Times*, September 16, 1979.

11. In February 1982, Andersen, who had left Smith's Rhodesian Front party and sat in parliament as an independent, was named minister of the public service in Mugabe's cabinet.

12. Carrington raised the issue of the visit of the two staffers with Secretary of State Vance, who asked the help of the Senate leadership to have the visitors leave London. The leadership asked Vance to obtain a written statement from Carrington to help them in their

approach to Helms. Carrington, for his own domestic political reasons, could not oblige. Thus, when the news leaked, Senator Helms told the press that he had been assured by the British that there had been no formal protest. The episode was embarrassing all around, but the staffers soon departed from London. For Vance's account see *Hard Choices*, p. 300.

13. *Financial Times*, September 19, 1979.
14. Jaster, p. 15.
15. *Cmnd.* 7802, pp. 17–33.
16. *Guardian*, October 6, 1979.
17. *JSAA* 4(4), pp. 490–493.
18. *Guardian*, October 11, 1979.
19. *Guardian*, October 10, 1979.
20. *Guardian*, October 16, 1979.
21. *Guardian*, October 16, 1979.
22. The exact wording, as repeated in the State Department press briefings, was: "We believe a multidonor effort would be appropriate to assist in the agricultural and economic development of an independent Zimbabwe within the framework of a wider development concept for southern Africa as a whole and we would be prepared to cooperate in such an effort."
23. *Time*, October 29, 1979.
24. *Financial Times*, October 19, 1979.

Chapter 5: The Transition

1. The text of the plan appears in *JSAA* 4(4), pp. 501–503.
2. Renwick, "The Rhodesian Settlement."
3. *Financial Times*, October 23, 1979.
4. "Patriotic Front Analysis of British Proposals for Interim Period," *JSAA* 4(4), pp. 498–500.
5. Walls had been following the conference closely from Salisbury and relaying advice long distance. On October 6 the *Guardian* reported that three weeks earlier Walls had sent a message to Muzorewa to accept the British constitutional proposals and agree to new elections.
6. *Financial Times*, October 29, 1979.
7. *Cmnd.* 7802, pp. 34–39.
8. *Financial Times*, November 8, 1979.
9. Colin Legum and David Martin in the *Observer*, November 11, 1979.
10. *Financial Times*, November 14, 1979.

11. In an exposé of in-country spying by British intelligence agencies, the *Times* reported on February 3, 1980, that, "by far the biggest intelligence sweep in recent years came at the prolonged Lancaster House Conference to thrash out the future of Zimbabwe-Rhodesia. Bluntly, the British Government wanted to know as much as possible about the tactics and goals of each delegation—and sanctioned a massive operation to find out. Phones were tapped. Hotel rooms were bugged. Diplomatic communications were monitored. 'That was why Lord Carrington could con the Conference on the basis of brinkmanship,' the source said. 'The intelligence services told him where the brinks were.' A particular target was the Patriotic Front leaders, Joshua Nkomo and Robert Mugabe."

12. According to the account in Martin Meredith's *The Past Is Another Country,* pp. 383–384.

13. *Cmnd.* 7802, p. 35.

14. *New York Times,* November 16, 1979.

15. Ibid.

Chapter 6: The Ceasefire

1. Martin and Johnson, p. 318.
2. Vance, p. 300.
3. *Financial Times,* November 20, 1979.
4. *Financial Times,* November 24, 1979.
5. *Financial Times,* December 1, 1979.
6. *Observer,* December 2, 1979.
7. *Guardian,* December 4, 1979.
8. *Time,* December 17, 1979.
9. *Guardian,* December 6, 1979.
10. Ibid.
11. Ibid.
12. *Guardian,* December 7, 1979.
13. Despite the fact that the transitional period agreement had specifically said, "The Governor will proceed to Rhodesia as soon as possible *after* the conclusion of the Constitutional Conference." *Cmnd.* 7802, p. 32.
14. *Cmnd.* 7802, p. 48.
15. *Cmnd.* 7802, p. 50.
16. *Financial Times,* December 15, 1979.
17. *Cmnd.* 7802, p. 56.

18. Vance, p. 301. Actually, no voter registration took place. All adults with a claim to Zimbabwean citizenship were allowed to vote without prior registration, which would have been a lengthy process.

19. U.S. Congress, House Committee on Foreign Affairs, *Executive-Legislative Consultation on Foreign Policy—Sanctions Against Rhodesia*, by Raymond Copson (Washington, D.C.: U.S. Government Printing Office, 1982), pp. 67–68.

20. *Guardian*, December 19, 1979.

21. John Womack, *Zapata and the Mexican Revolution* (New York: Alfred A. Knopf, 1969), p. 205.

22. Concluding the conference with Mugabe's acceptance of the British offer of an additional assembly point, for a total of sixteen, constituted an almost classic example of Schelling's concept of a "focal point," that is, "a mutually identifiable resting place . . . [that] may not be so much conspicuously fair or conspicuously in balance with estimated bargaining powers as just plain 'conspicuous.' " Focal points provide negotiators with conveniently salient features toward which bargaining can be directed. Cross argues that focal point acceptance constitutes a form of arbitration (in which the parties are complicit) that "substitutes for what would otherwise be a very expensive debate." The sixteenth assembly point, which closed the conference, was, of course, not the only example of the utilization of focal points by the parties during the Lancaster House Conference. The pledge on land redistribution at the end of the constitutional discussions, and the alteration of the U.K.'s forty-one points, which allowed the transition debate to come to an end, were two of the more prominent of the other instances. (See Schelling, *The Strategy of Conflict*, pp. 67–74, and Cross, *The Economics of Bargaining*, pp. 92–97.)

Chapter 7: Denouement

1. Meredith, p. 393.
2. Ibid., pp. 396–397.

Chapter 8: Accounting for Success

1. See, for instance, Callinicos, pp. 51–71.

2. Dean G. Pruitt makes this point about Kissinger's Middle Eastern negotiating tactics in "Kissinger as a Traditional Mediator with Power,"

in Jeffrey Rubin, *Dynamics of Third Party Intervention* (New York: Praeger, 1981).

3. Newhouse, p. 73.

4. *Time,* October 8, 1979.

5. Fred C. Ikle, *How Nations Negotiate* (New York: Harper & Row, 1965), p. 2.

6. For an interesting comparison of Rhodesia and Angola as viewed by the United States and the USSR, see Larry C. Napper, "The African Terrain and U.S.-Soviet Conflict in Angola and Rhodesia: Some Implications for Crisis Prevention," in Alexander L. George, *Managing U.S.-Soviet Rivalry* (Boulder: Westview Press, 1983).

7. See Martin and Johnson, *The Struggle for Zimbabwe,* pp. 305–308, for a curious attempt by Cuba, shortly before the conference, to internationalize the conflict.

8. Sir William Hayter, *The Diplomacy of Great Powers* (New York: Macmillan, 1961), pp. 48–49.

9. Sir Harold Nicolson, *Diplomacy* (London: Oxford University Press, 2nd ed., 1950), p. 132.

10. A point made by William Zartman in *The 50% Solution* (New York: Anchor Books, 1976).

11. *Financial Times,* December 19, 1979.

12. *Observer,* October 18, 1979.

13. Ikle, p. 26.

14. *Guardian,* October 9, 1979.

15. See, for instance, Callinicos, p. 50.

16. Ottomar J. Bartos, "How Predictable are Negotiations?" *Journal of Conflict Resolution* 11(4), December 1976.

17. See, for instance, Oran Young, *The Intermediaries* (Princeton: Princeton University Press, 1967), and F. S. Northedge and M. D. Donelan, *International Disputes* (London: Europa Publications, 1971).

18. Two recent volumes have explored the role of the biased mediator and provide significant insight. See Saadia Touval, *The Peace Brokers* (Princeton: Princeton University Press, 1982); and Jeffrey Z. Rubin, *Dynamics of Third Party Intervention* (New York: Praeger, 1981).

19. Roger Fisher, "Fractionating Conflict," in R. Fisher, ed., *International Conflict and Behavioral Science* (New York: Basic Books, 1964).

20. *Guardian,* October 8, 1979.

21. *Guardian,* October 10, 1979.

22. Vance, p. 299.

23. In November 1973, Egyptian General Gamasy and Israeli General Yariv began to negotiate a Sinai disengagement directly. Kissinger reacted negatively to negotiations in which the United States was not the controlling actor. He therefore imposed upon the parties to squash the direct talks.

24. The single-negotiating-text strategy is a device advocated by Roger Fisher. It was used successfully by President Carter at the Camp David meeting with Sadat and Begin. See Roger Fisher, "Playing the Wrong Game," in Rubin, *Dynamics of Third Party Intervention.*

25. "Huff and puff and fudge" is a phrase too apt not to appropriate from the *Economist,* October 13, 1979.

26. Thomas Schelling, *Arms and Influence* (Cambridge: Harvard University Press, 1960), pp. 69–78.

27. Roger Fisher and William Ury, *Getting to Yes* (Boston: Houghton Mifflin, 1981), pp. 101–111.

28. Shakespeare, *Hamlet,* Act IV, scene 5. Gentleman describing to the queen the reaction of others upon meeting the mad Ophelia.

Chapter 9: Dominant Third-Party Mediation

1. Ikle, p. 1.

2. Ikle talks about the negotiator's tendency to mediation, pp. 143–149. The concept of "principled negotiation" is developed by Fisher and Ury in *Getting to Yes.*

3. "Mediation with muscle" is a term developed by Donald B. Straus in "Kissinger and the Management of Complexity: An Attempt that Failed," in Rubin's *Dynamics of Third Party Intervention.*

4. Arbitration is another form of conflict resolution involving a third party. Its use in international affairs is, however, limited because the arbitrator generally has little or no power, apart from moral suasion, to oblige the disputants to accept his decision. A variation of arbitration that is becoming more common in labor disputes is termed med-arb. In it the parties agree that if a third party is unsuccessful in mediating a dispute, he may impose an arbitrated solution. In some ways med-arb resembles what Carrington did at Lancaster House. However, the important distinction is that med-arb, like straight arbitration, requires the disputants to enter into a prior agreement to accept the third party's determinations. For this reason, med-arb is as likely to be of limited utility in international application as is the more traditional form of arbitration.

5. Stephen Low, "The Zimbabwe Settlement," unpublished, 1982.

Selected Bibliography

A note on periodicals: The *Times* was on strike for much of the period of the conference. Of British newspapers, the *Financial Times* provided the most useful day-to-day coverage. The *Guardian* and weekly *Observer* were also helpful. R. W. Apple's reporting in the *New York Times*, though less frequent than his British colleagues', was often more informative. Of the weekly news magazines, the *Economist* was often too involved in sermonizing to be of great use. *Time* and *Newsweek* were occasionally of value. The best summaries of the conference's proceedings were provided in Michael Clough's articles in *Africa Index*. The collection of documents in the *Journal of Southern African Affairs*, vol. 4, no. 4, was of some use in compensating for the dearth of published documentation but must be used with care to compensate for the absence of dates and its incompleteness.

Books and Articles

Bartos, Ottomar. *Process and Outcome Negotiations*. New York: Columbia University Press, 1974.

————. "How Predictable are Negotiations?" *Journal of Conflict Resolution* 11(4), December 1976, 481–517.

Brzezinski, Zbigniew. *Power and Principle*. New York: Farrar, Straus & Giroux, 1981.

Callinicos, Alex. *Southern Africa After Zimbabwe*. London: Pluto Press, 1981.

Caplow, Theodore. *Two Against One*. Englewood Cliffs: Prentice-Hall, 1968.

Clough, Michael, ed. *Changing Realities in Southern Africa*. Berkeley: Institute of International Studies, 1982.

Cross, John G. *The Economics of Bargaining.* New York: Basic Books, 1969.

Diesing, Paul and Snyder, Glenn. *Conflict Among Nations.* Princeton: Princeton University Press, 1971.

Donelan, M. D. and Northedge, F. S. *International Disputes.* London: Europa Publications, 1971.

Fisher, Roger. *International Conflict for Beginners.* New York: Harper & Row, 1969.

——————, ed. *International Conflict and Behavioral Science.* New York: Basic Books, 1964.

——————, and Ury, William. *Getting to Yes.* Boston: Houghton-Mifflin, 1981.

George, Alexander L. *Managing U.S.-Soviet Rivalry.* Boulder: Westview Press, 1983.

Great Britain. Secretary of State for Foreign and Commonwealth Affairs. *Report of the Constitutional Conference.* London: Her Majesty's Stationery Office, Cmnd. 7802, 1980.

Handel, Michael. "Surprise and Change in International Politics." *Journal of International Security* 4(4), Spring 1980, 57–85.

Harriman, W. Averell. "Observations on Negotiating: Informal Views of W. Averell Harriman." *Journal of International Affairs* 29(1), Spring 1975, 1–6.

Hayter, Sir William. *The Diplomacy of Great Powers.* New York: Macmillan, 1961.

Hills, Denis. *The Last Days of White Rhodesia.* London: Chatto and Windus, 1981.

Hudson, Miles. *Triumph or Tragedy?* London: Hamish-Hamilton, 1981.

Ikle, Fred Charles. *How Nations Negotiate.* New York: Harper & Row, 1965.

Jaster, Robert. *A Regional Security Role for Africa's Front-Line States: Experience and Prospects.* London: International Institute for Strategic Studies (Adelphi Paper 180), 1983.

Kelman, Herbert, ed. *International Behavior.* New York: Holt, Rinehart and Winston, 1965.

Low, Amb. Stephen. "The Zimbabwe Settlement." (unpublished), 1982.

Martin, David and Johnson, Phyllis. *The Struggle for Zimbabwe.* London: Faber and Faber, 1981.

Meredith, Martin. *The Past Is Another Country.* London: Pan Books, 1980.

Mitchell, Diana. *African Nationalist Leaders in Zimbabwe: Who's Who 1980.* Salisbury: Diana Mitchell, 1980.

Naipaul, Shiva. *North of South.* London: Penguin Books, 1980.

Newhouse, John. "Profiles (Lord Carrington)." *New Yorker,* February 14, 1983.

Nicolson, Sir Harold. *Diplomacy.* London: Oxford University Press, 2nd ed., 1950.

O'Meara, Patrick and Carter, Gwendolyn, eds. *Southern Africa: The Continuing Crisis.* Bloomington: Indiana University Press, 2nd ed., 1982.

Peele, Gillian. "The Changed Character of British Foreign and Security Policy." *Journal of International Security* 4(4), Spring 1980, 185–198.

Quandt, William. "Kissinger and the Arab-Israeli Disengagement Negotiations." *Journal of International Affairs* 29(1), Spring 1975, 33–48.

Renwick, Robin. *Economic Sanctions.* Cambridge: Harvard University, Center for International Affairs, 1981.

_____ . "The Rhodesia Settlement" (unpublished). Cambridge: Harvard University, Center for International Affairs, 1981.

Rotberg, Robert. *Suffer the Future.* Cambridge: Harvard University Press, 1980.

Rubin, Jeffrey Z., ed. *Dynamics of Third Party Intervention.* New York: Praeger, 1981.

_____ , and Brown, Bert B. *The Social Psychology of Bargaining and Negotiation.* New York: Academic Press, 1975.

Sanger, Clyde, "Zimbabwe: A New Beginning After 14 Years of Ian Smith." *International Perspective,* January/February 1980, 14–17.

Schelling, Thomas. *The Strategy of Conflict.* Cambridge: Harvard University Press, 1960.

_____ . *Arms and Influence.* New Haven: Yale University Press, 1966.

Smiley, Xan. "Zimbabwe, Southern Africa and the Rise of Robert Mugabe." *Foreign Affairs* 58(5), Summer 1980, 1060–1083.

Soames, Lord Christopher. "Rhodesia to Zimbabwe." *International Affairs* 56(3), Summer 1980, 405–419.

Touval, Saadia. *The Peace Brokers.* Princeton: Princeton University Press, 1982.

U.S. Congress, House Committee on Foreign Affairs. *Executive-Legislative Consultation on Foreign Policy: Sanctions Against Rhodesia* (by

Raymond Copson). Washington: U.S. Government Printing Office (no. 97-5590), 1982.

Vance, Cyrus. *Hard Choices.* New York: Simon and Schuster, 1983.

Waltz, Kenneth. *Foreign Policy and Democratic Politics: The American and British Experiences.* Boston: Little, Brown, 1967.

Womack, John. *Zapata and the Mexican Revolution.* New York: Alfred A. Knopf, 1969.

Young, Oran. *The Intermediaries.* Princeton: Princeton University Press, 1967.

————— . "Intermediaries: Additional Thoughts on Third Parties." *Journal of Conflict Resolution* 16(1), March 1972, 51–66.

Zartman, I. William. "The Political Analysis of Negotiation: How Who Gets What and When." *World Politics* 26(3), April 1974, 385–399.

————— . "Negotiations: Theory and Reality." *Journal of International Affairs* 29(1), Spring 1975, 60–78.

————— . *The 50% Solution.* New York: Anchor Books, 1976.

————— , ed. *The Negotiation Process.* Beverly Hills: Sage Publications, 1978.

Index